CHM-101 General Chemistry (I) Laboratory Manual

Author

Hein

Printed in the United States of America 10 9 8 7 6 5 4 3 2

List of Titles

Introduction to General, Organic, and Biochemistry in the Laboratory, 10th edition
by Morris Hein, Judith N. Peisen, and James M. Ritchey
 ISBN: 978-0-470-59881-8

Table of Contents

Chapter 1. CHEMISTRY DEPARTMENT SAFETY PROCEDURES 041510 7

Original material provided by the instructors

Experiment 2: Measurements 10

Originally from *Introduction to General, Organic, and Biochemistry in the Laboratory, 10th edition*

Blank Notes Page 23

Experiment 5: Calorimetry and Specific Heat 24

Originally from *Introduction to General, Organic, and Biochemistry in the Laboratory, 10th edition*

Experiment 6: Freezing Points—Graphing of Data 32

Originally from *Introduction to General, Organic, and Biochemistry in the Laboratory, 10th edition*

Experiment 7: Water in Hydrates 42

Originally from *Introduction to General, Organic, and Biochemistry in the Laboratory, 10th edition*

Blank Notes Page 49

Experiment 9: Properties of Solutions 50

Originally from *Introduction to General, Organic, and Biochemistry in the Laboratory, 10th edition*

Blank Notes Page 61

Experiment 10: Composition of Potassium Chlorate 62

Originally from *Introduction to General, Organic, and Biochemistry in the Laboratory, 10th edition*

Blank Notes Page 69

Experiment 13: Ionization-Electrolytes and pH 70

Originally from *Introduction to General, Organic, and Biochemistry in the Laboratory, 10th edition*

Blank Notes Page 85

Experiment 14: Identification of Selected Anions 86

Originally from *Introduction to General, Organic, and Biochemistry in the Laboratory, 10th edition*

Experiment 17: Lewis Structures and Molecular Models 94

Originally from *Introduction to General, Organic, and Biochemistry in the Laboratory, 10th edition*

Blank Notes Page 107

X Experiment 22: Neutralization–Titration I 108

for Dec. 4th

Originally from *Introduction to General, Organic, and Biochemistry in the Laboratory, 10th edition*
Blank Notes Page 117
Experiment 23: Neutralization–Titration II 118
Originally from *Introduction to General, Organic, and Biochemistry in the Laboratory, 10th edition*
Blank Notes Page 127
Experiment 24: Chemical Equilibrium-Reversible Reactions 128
Originally from *Introduction to General, Organic, and Biochemistry in the Laboratory, 10th edition*
Chapter 2. MOLECULAR MASS OF AN IDEAL GAS 137
Original material provided by the instructors

EXPERIMENT 2

Measurements

MATERIALS AND EQUIPMENT

Solids: sodium chloride (NaCl) and ice. Balance, ruler, thermometer, solid object for density determination, No. 1 or 2 solid rubber stopper.

DISCUSSION

Chemistry is an experimental science, and measurements are fundamental to most of the experiments. It is important to learn how to make and use these measurements properly.

The SI System of Units

The International System of Units (*Systeme Internationale, SI*) or metric system is a decimal system of units for measurements used almost exclusively in science. It is built around a set of units including the meter, the gram, and the liter and uses factors of 10 to express larger or smaller multiples of these units. To express larger or smaller units, prefixes are added to the names of the units. Deci, centi, and milli are units that are 1/10, 1/100, and 1/1000, respectively, of these units. The most common of these prefixes with their corresponding values expressed as decimals and powers of 10 are shown in the table below.

Prefix	Decimal Equivalent	Power of 10	Examples
deci (d)	0.1	10^{-1}	1 dg = 0.1 g = 10^{-1} g
centi (c)	0.01	10^{-2}	1 cm = 0.01 m = 10^{-2} m
milli (m)	0.001	10^{-3}	1 mg = 0.001 g = 10^{-3} g
kilo (k)	1000	10^{3}	1 km = 1000 m = 10^{3} m

Dimensional Analysis

It will often be necessary to convert from the American System of units to the SI system or to convert units within the SI system. Conversion factors are available from tables (see Appendix 4) or can be developed from the metric prefixes and their corresponding values as shown in the table above. Dimensional analysis, a problem-solving method with many applications in chemistry, is very valuable for converting one unit to another by the use of conversion factors. A review of using dimensional analysis for converting units is provided here. Study Aid 5 provides more help with this problem-solving tool.

Conversion Factors come from equivalent relationships, usually stated as equations. From each equivalence statement two conversion factors can be written in fractional form with a value of 1. For example:

Equivalence Equations	Conversion Factor #1	Conversion Factor #2
1 dollar = 4 quarters	$\frac{1\text{ dollar}}{4\text{ quarters}}$	$\frac{4\text{ quarters}}{1\text{ dollar}}$
1 lb = 453.6 g	$\frac{1\text{ lb}}{453.6\text{ g}}$	$\frac{453.6\text{ g}}{1\text{ lb}}$
1 mm = 10^{-3} m	$\frac{1\text{ mm}}{10^{-3}\text{m}}$	$\frac{10^{-3}\text{m}}{1\text{ mm}}$
1 ns = 10^{-9} s	$\frac{1\text{ ns}}{10^{-9}\text{ s}}$	$\frac{10^{-9}\text{ s}}{1\text{ ns}}$

The dimensional analysis method of converting units involves organizing one or more conversion factors into a logical series which cancels or eliminates all units except the unit(s) wanted in the answer.

For example: To convert 2.53 lb into milligrams (mg), the setup is:

$$(2.53\text{ lb})\left(\frac{453.6\text{ g}}{1\text{ lb}}\right)\left(\frac{1\text{ mg}}{10^{-3}\text{g}}\right) = 1.15 \times 10^{6}\text{ mg}$$

Note, that in completing this calculation, units are treated as numbers, **lb** in the denominator is canceled into **lb** in the numerator and **g** in the denominator is cancelled into **g** in the numerator. More examples of unit conversions can be found in Study Aid 5.

Although the SI unit of temperature is the Kelvin (K), the Celsius (or centigrade) temperature scale is commonly used in scientific work and the Fahrenheit scale is commonly used in this country. On the Celsius scale the freezing point of water is designated 0°C, the boiling point 100°C.

Precision and Accuracy of Measurements

Scientific measurements must be as **precise** as possible. This means that every measurement will include one uncertain or estimated digit. When making measurements we normally estimate between the smallest scale divisions on the instrument being used. Then, only the uncertain digit should vary if the measurement is repeated using the same instrument, even if it is repeated by someone else. The **accuracy** of a measurement or calculated quantity refers to its agreement with some known value. For example, we need to make two measurements, volume and mass, to determine the density of a metal. This experimental density can then be compared with the density of the metal listed in a reference such as the *Handbook of Chemistry and Physics.* High accuracy means there is good agreement between the experimental value and the known value listed in the reference. Not all measurements can be compared with a known value.

Random and Systematic Errors

The difference between the experimentally measured value of something and the accepted value of something is known as **the error.** For many of the experiments in this course, after you determine the error in your result, you may be required to find the percent error:

$$\text{Percent error} = \frac{\text{theoretical accepted value} - \text{experimentally determined value}}{\text{theoretical accepted value}} \times 100$$

There are two different types of error. **A random error** means that the error has an equal probablilty of being higher or lower than the accepted value. For example, a student measures the density of a quartz sample four times: (Accepted density value for quartz is 2.65 g/mL)

	2.72 g/mL	
	2.55 g/mL	Since two of the measured density values are below the mean
	2.68 g/mL	and two are above the mean, there is an **equal probability** of the
	2.60 g/mL	measurements being above or below the mean. This is a **random** error.
		Since the mean density value is very close to the accepted value, the
Mean =	2.64 g/mL	accuracy of the mean measurement is good. (the percent error is 0.38%)

The other type of error is a **systematic error.** This type of error occurs in the same direction each time (either always higher or always lower than the accepted value). For example, a student measures the boiling point of water four times (accepted temperature for the boiling point of water is 100.0° C.)

	101.2° C	
	100.9° C	Since all four of the measured temperature values are above the accepted
	102.0° C	value, the **error** is systematic. The mean value is 1.3% higher than the
	101.0° C	accepted value so the accuracy of these measurements is not as good as the
Mean =	101.3° C	accuracy of the density of the measurements in the first example.

Precision and Significant Figures

When a measured value is determined to the highest precision of the measuring instrument, the digits in the measurement are called **significant digits** or **significant figures.**

Suppose we are measuring two pieces of wire, using the metric scale on a ruler that is calibrated in tenths of centimeters as shown in Figures 2.1a and b. One end of the first wire is placed at exactly 0.0 cm and the other end falls somewhere between 6.3 cm and 6.4 cm. Since the distance between 6.3 and 6.4 is very small, it is difficult to determine the next digit exactly. One person might estimate the length of the wire as 6.34 cm and another as 6.33 cm. The estimated digit is never ignored because it tells us that the ruler can be read to the 0.01 place. This measurement therefore has three significant figures (two certain and one uncertain figure).

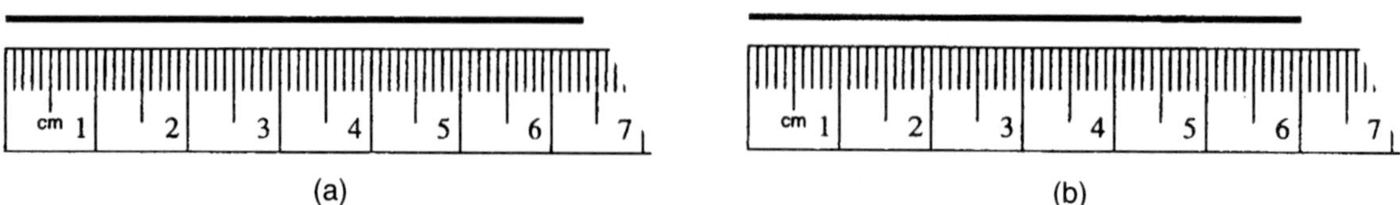

(a) (b)

Figure 2.1

The second wire has a length which measures exactly 6 cm on the ruler as shown in Figure 2.1b. Reporting this length as 6 cm would be a mistake for it would imply that the 6 is an uncertain digit and others might record 5 or 7 as the measurement. Recording the measurement as 6.0 would also be incorrect because it implies that the 0 is uncertain and that someone else might estimate the length as 6.1 or 5.9. What we really mean is that, as closely as we can read it, the length is exactly 6 cm. So, we must write the number in such a way that it tells how precisely we can read it. In this example we can estimate to 0.01 cm so the length should be reported as 6.00 cm.

Significant Figures in Calculations

The result of multiplication, division, or other mathematical manipulation cannot be more precise than the least precise measurement used in the calculation. For instance, suppose we have an object that weighs 3.62 lb and we want to calculate the mass in grams. $(3.62 \text{ lb})\left(\frac{453.6 \text{ g}}{1 \text{ lb}}\right) = 1{,}642.032$ when done by a calculator. To report 1,642.032 g as the mass is absurd, for it implies a precision far beyond that of the original measurement. Although the conversion factor has four significant figures, the mass in pounds has only three significant figures. Therefore the answer should have only three significant figures; that is, 1,640 g. In this case the zero cannot be considered significant. This value can be more properly expressed as 1.64×10^3g. For a more comprehensive discussion of significant figures see Study Aid 1.

Precise Quantities versus Approximate Quantities

In conducting an experiment it is often unnecessary to measure an exact quantity of material. For instance, the directions might state, "Weigh about 2 g of sodium sulfite." This instruction indicates that the measured quantity of salt should be 2 g plus or minus a small quantity. In this example 1.8 to 2.2 g will satisfy these requirements. To weigh exactly 2.00 g or 2.000 g wastes time since the directions call for approximately 2 g.

Sometimes it is necessary to measure an amount of material precisely within a stated quantity range. Suppose the directions read, "weigh about 2 g of sodium sulfite to the nearest 0.001 g." This instruction does not imply that the amount is 2.000 g but that it should be between 1.8 and 2.2 g and measured and recorded to three decimal places. Therefore, four different students might weigh their samples and obtain 2.141 g, 2.034 g, 1.812 g, and 1.937 g, respectively, and each would have satisfactorily followed the directions.

Temperature

The simple act of measuring a temperature with a thermometer can easily involve errors. Not only does the calibration of the scale on the thermometer limit the precision of the measurement, but the improper placement of the thermometer bulb in the material being measured introduces a common source of human error. When measuring the temperature of a liquid, one can minimize this type of error by observing the following procedures:

1. Hold the thermometer away from the walls of the container.
2. Allow sufficient time for the thermometer to reach equilibrium with the liquid.
3. Be sure the liquid is adequately mixed.

When converting from degrees Celsius to Fahrenheit or vice versa, we make use of the following formulas:

$$°C = \frac{(°F - 32)}{1.8} \quad \text{or} \quad °F = (1.8 \times °C) + 32$$

Example Problem: Convert 70.0°F to degrees Celsius:

$$°C = \left(\frac{70.0°F - 32}{1.8}\right) = \frac{38.0}{1.8} = 21.11°C \text{ rounded to } 21.1°C$$

This example shows not only how the formula is used but also a typical setup of the way chemistry problems should be written. It shows how the numbers are used, but does not show the multiplication and division, which should be worked out by calculator. The answer was changed from 21.11°C to 21.1°C because the initial temperature, 70.0°F, has only three significant figures. The 1.8 and 32 in the formulas are exact numbers and have no effect on the number of significant figures.

Mass (Weight)

The directions in this manual are written for a 0.001 gram precision balance, but all the experiments can be performed satisfactorily using a 0.01 gram or 0.0001 gram precision balance. Your instructor will give specific directions on how to use the balance, but the following precautions should be observed:

1. The balance should always be "zeroed" before anything is placed on the balance pan. On an electronic digital balance, this is done with the "tare" or "T" button. Balances without this feature should be adjusted by the instructor.

2. Never place chemicals directly on the balance pan; first place them on a weighing paper, weighing "boat", or in a container. Clean up any materials you spill on or around the balance.

3. Before moving objects on and off the pan, be sure the balance is in the "arrest" position. When you leave the balance, return the balance to the "arrest" or standby position.

4. Never try to make adjustments on a balance. If it seems out of order, tell your instructor.

Volume

Beakers and flasks are marked to indicate only approximate volumes. Volume measurements are therefore made in a graduated cylinder by reading the point on the graduated scale that coincides with the bottom of the curved surface called the **meniscus** of the liquid (Figure 2.2). Volumes measured in this illustrated graduated cylinder are calibrated in 1 mL increments and should be estimated and recorded to the nearest 0.1 mL.

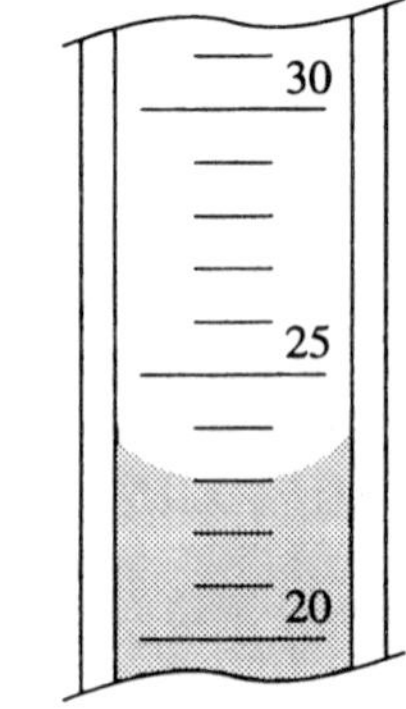

Figure 2.2 Read the bottom of the meniscus. The volume is 23.0 mL

Density

Density is a physical property of a substance and is useful in identifying the substance. **Density** is the ratio of the mass of a substance to the volume occupied by that mass; it is the mass per unit volume and is given by the equations

$$\text{Density} = d = \frac{\text{Mass}}{\text{Volume}} = \frac{m}{V} = \frac{g}{mL} \text{ or } \frac{g}{cm^3}$$

In calculating density it is important to make correct use of units and mathematical setups.

Example Problem: An object weighs 283.5 g and occupies a volume of 14.6 mL. What is its density?

$$d = \frac{m}{V} = \frac{283.5 \text{ g}}{14.6 \text{ mL}} = 19.4 \text{ g/mL}$$

Note that all the operations involved in the calculation are properly indicated and that all units are shown. If we divide grams by milliliters, we get an answer in grams per milliliter.

The volume of an irregularly shaped object is usually measured by the displacement of a liquid. An object completely submerged in a liquid displaces a volume of the liquid equal to the volume of the object.

Measurement data and calculations must always be accompanied by appropriate units.

PROCEDURE

Wear protective glasses.

Record your data on the report form as you complete each measurement, never on a scrap of paper which can be lost or misplaced.

A. Temperature

Record all temperatures to the **nearest 0.1°C.**

1. Fill a 400 mL beaker half full of tap water. Place your thermometer in the beaker. Give it a minute to reach thermal equilibrium. Keeping the thermometer in the water and holding the tip of the thermometer away from the glass, read and record the temperature.

2. Fill a 150 mL beaker half full of tap water. Set up a ring stand with the ring and wire gauze at a height so the hottest part of the burner flame will reach the bottom of the beaker. Heat the water to boiling. Read and record the temperature of the boiling water, being sure to hold the thermometer away from the bottom of the beaker.

3. Fill a 250 mL beaker one-fourth full of tap water and add a 100 mL beaker of crushed ice. Without stirring, place the thermometer in the beaker, resting it on the bottom. Wait at least 1 minute, then read and record the temperature. Now stir the mixture for about 1 minute. If almost all the ice melts, add more. Holding the thermometer off the bottom, read and record the temperature. Save the ice-water mixture for Part 4.

4. Weigh approximately 5 g of sodium chloride and add it to the ice-water mixture. Stir for 1 minute, adding more ice if needed. Read and record the temperature. Dispose of the salt water/ice mixture in the sink.

WASTE DISPOSE OF PROPERLY

B. Mass

Using the balance provided, do the following, recording all the masses to include one uncertain digit and all certain digits.

1. Weigh a 250 mL beaker.

2. Weigh a 125 mL Erlenmeyer flask.

3. Weigh a piece of weighing paper or a plastic weighing "boat."

4. Add approximately 2 g of sodium chloride to the weighing paper from step 3 and record the total mass. Calculate the mass of sodium chloride.

C. Length

Using a ruler, make the following measurements in centimeters; measure to the nearest uncertain digit.

1. Measure the length of the arrow on the right ⟶

2. Measure the external height of a 250 mL beaker.

3. Measure the length of a test tube.

D. Volume

Using the graduated cylinder most appropriate, measure the following volumes to the maximum precision possible, usually 0.1 mL. Remember to read the volume at the meniscus.

1. Fill a test tube to the brim with water and measure the volume of the water.

2. Fill a 125 mL Erlenmeyer flask to the brim with water and measure the volume of the water.

3. Measure 5.0 mL of water in a graduated cylinder and pour it into a test tube. With a ruler, measure the height (in cm) and mark the height with a marker.

4. Measure 10.0 mL of water in the graduated cylinder and pour it into a test tube like the one used in the previous step. Again, mark the height with a marker.

In the future, you will often find it convenient to estimate volumes of 5 and 10 mL simply by observing the height of the liquid in the test tube.

E. Density

Estimate and record all volumes to the highest precision, usually 0.1 mL. Make all weighing to the highest precision of the balance. Note that you must supply the units for the measurements and calculations in this section.

1. Density of Water. Weigh a clean, dry 50 mL graduated cylinder and record its mass. (Graduated cylinders should never be dried over a flame.) Fill the graduated cylinder with distilled water to 50.0 mL. Use a medicine dropper to adjust the meniscus to the 50.0 mL mark. Record the volume. Reweigh and calculate the density of water.

2. Density of a Rubber Stopper. Select a solid rubber stopper which is small enough to fit inside the 50 mL graduated cylinder. Weigh the dry stopper. Fill the 50 mL cylinder with tap water to approximately 25 mL. Read and record the exact volume. Carefully place the rubber stopper into the graduated cylinder so that it is submerged. Read and record the new volume. Calculate the volume and density of the rubber stopper.

3. Density of a Solid Object. Obtain a solid object from your instructor. Record the sample code on the report form. Determine the density of your solid by following the procedure given in Part 2 for the rubber stopper. To avoid the possibility of breakage, incline the graduated cylinder at an angle and slide, rather than drop, the solid into it.

Return the solid object to your instructor.

NAME ______________________

SECTION __________ DATE __________

REPORT FOR EXPERIMENT 2

INSTRUCTOR ______________________

Measurements

A. Temperature

1. Water at room temperature ______________ °C
2. Boiling point ______________ °C
3. Ice water

   Before stirring ______________ °C

   After stirring for 1 minute ______________ °C
4. Ice water with salt added ______________ °C

B. Mass

1. 250 mL beaker ______________ g
2. 125 mL Erlenmeyer flask ______________ g
3. Weighing paper or weighing boat ______________ g
4. Mass of weighing paper/boat + sodium chloride ______________ g

   Mass of sodium chloride (show calculation setup) ______________ g

C. Length

1. Length of ——————————→ ______________ cm
2. Height of 250 mL beaker ______________ cm
3. Length of test tube ______________ cm

D. Volume

1. Test tube ______________ mL
2. 125 mL Erlenmeyer flask ______________ mL
3. Height of 5.0 mL of water in test tube ______________ cm
4. Height of 10.0 mL of water in test tube ______________ cm

 NAME ______________________

E. Density

1. Density of Water

Mass of empty graduated cylinder ______________

Volume of water ______________

Mass of graduated cylinder and water ______________

Mass of water (show calculation setup) ______________

Density of water (show calculation setup) ______________

2. Density of a Rubber Stopper

Mass of rubber stopper ______________

Initial volume of water in cylinder ______________

Final volume of water in cylinder (including stopper) ______________

Volume of rubber stopper (show calculation setup) ______________

Density of rubber stopper (show calculation setup) ______________

3. Density of a Solid Object

Number of solid object ______________

Mass of solid object ______________

Initial volume of water in graduated cylinder ______________

Final volume in graduated cylinder ______________

Volume of solid object (show calculation setup) ______________

Density of solid object (show calculation setup) ______________

QUESTIONS AND PROBLEMS

1. The directions state "weigh about 5 grams of sodium chloride". Give minimum and maximum amounts of sodium chloride that would satisfy these instructions.

2. Two students each measured the density of a quartz sample three times:

	Student A	*Student B*
1.	3.20 g/mL	2.82 g/mL
2.	2.58 g/mL	2.48 g/mL
3.	2.10 g/mL	2.59 g/mL
mean	2.63 g/mL	2.63 g/mL

The density found in the *Handbook of Chemistry and Physics* for quartz is 2.65 g/mL

(a) Which student measured density with the greatest precision? Explain your answer.

(b) Which student measured density with the greatest accuracy? Explain your answer.

(c) Are the errors for these students random or systematic? Explain.

Show calculation setups and answers for the following problems.

3. Convert 21°C to degrees Fahrenheit. ____________

4. Convert 101°F to degrees Celsius. ____________

5. An object is 9.6 cm long. What is the length in inches? ____________

6. An empty graduated cylinder weighs 82.450 g. When filled to 50.0 mL with an unknown liquid it weighs 110.810 g. What is the density of the unknown liquid?

____

7. It is valuable to know that 1 milliliter (mL) equals 1 cubic centimeter (cm^3 or cc). How many cubic centimeters are in an 8.00 oz bottle of cough medicine? (1.00 oz = 29.6 mL)

____

8. A metal sample weighs 56.8 g. How many ounces does this sample weigh? (1 lb = 16 oz)

____

9. Convert 15 nm into km.

____

EXPERIMENT 5

Calorimetry and Specific Heat

MATERIALS AND EQUIPMENT

Styrofoam cups, 6 oz; thermometers, metal samples, test tube with a diameter of at least 22 mm, bunsen burner, wire gauze, 400 ml beaker, cardboard cut into 4" squares with a small hole in the middle for a thermometer.

DISCUSSION

Calorimetry is the science of measuring a quantity of heat. Heat is a form of energy associated with the motion of atoms or molecules of a substance. Heat (often represented as "q") is measured in energy units such as joules or calories. Temperature (often represented as "t") is measured in degrees (usually Celsius). The measurement of temperature is already familiar to you. The same temperature is obtained for the water in a lake and for a thermos of water taken from the lake. But the heat content of the whole lake is much more than the heat content in that thermos of water even though both are exactly the same temperature.

Temperature and heat are related to each other by the specific heat (*sp ht*) of a substance, defined as the quantity of heat needed to raise one gram of a substance by one degree Celsius (J/g°C). The relationship between quantity of heat (q), specific heat (*sp ht*), mass (m) and temperature change (Δt) is mathematically expressed by the equation:

$$q = (m)(sp\ ht)(\Delta t) \quad \text{or Joules} = (\text{g})\left(\frac{\text{J}}{\text{g}^\circ\text{C}}\right)(\Delta^\circ\text{C}) = (\cancel{\text{g}})\left(\frac{\text{J}}{\cancel{\text{g}^\circ\text{C}}}\right)(\Delta\cancel{^\circ\text{C}}) = \text{J}$$

Since the mass and temperature can be measured by a balance and a thermometer, respectively, q can be calculated if the *sp ht* for a substance is known. Also, sp *ht* can be calculated if the heat content (q) of the substance is known. The amount of heat needed to raise the temperature of 1 g of water by 1 degree Celsius is the basis of the calorie. Thus, the specific heat of water is exactly 1.00 cal/g°C. The SI unit of energy is the joule and it is related to the calorie by 1 calorie = 4.184 J. Thus, the specific heat of water is also 4.184 J/g°C. The specific heat of a substance relates to its capacity to absorb heat energy. **The higher the specific heat of a substance the more energy required to change its temperature.**

The specific heat of metals generally varies with their atomic masses. You will see this relationship later when the data in Table 5.1 is graphed. For this data, atomic mass is the independent variable and specific heat is the dependent variable because we are examining how specific heat changes as a function of atomic mass. For a review of variables and graphing techniques, see Study Aid 3.

In this experiment, we will use calorimetry to determine the specific heat of a metal. Heat energy is transferred from a hot metal to water until the metal and the water have reached the same temperature. This transfer is done in an insulated container to minimize heat losses to

Table 5.1 Specific Heat of Selected Metals

Name of Metal	Atomic Mass, amu	Specific Heat, J/g°C
Aluminum	26.98	0.900
Copper	63.55	0.385
Gold	197.0	0.131
Iron	55.85	0.451
Lead	207.2	0.128
Silver	107.9	0.237
Tin	118.7	0.222

the surroundings. We then make the assumption that all the heat lost by the metal (q_x) was absorbed by the water and is equal to the heat gained by the water, (q_w). Since we know the specific heat of water, we have all the variables needed to calculate q_w using the equation:

$$q_w = (m_w)(sp\ ht_w)(\Delta t_w)$$

Since q_w is equal to q_x we can say that

$$q_w = q_x = (m_x)(sp\,ht_x)(\Delta t_x)$$

This relationship can be used to calculate $sp\ ht_x$ of a metal because both m_x and Δt_x can be measured.

Sample Calculation:

A metal sample weighing 68.3820 g was heated to 99.0°C, then quickly transferred into a styro-foam calorimeter containing 62.5515 g of distilled water at a temperature of 18.0°C. The temperature of the water in the styrofoam cup increased and stabilized at 20.6°C. Calculate the $sp\ ht_x$ and identify the metal using 4.184 J/g°C for the $sp\ ht_w$.

$$\begin{aligned} \Delta t_w &= 20.6°\text{C} - 18.0°\text{C} = 2.6°\text{C} \\ q_w &= (m_w)(sp\,ht_w)(\Delta t_w) \\ &= (62.5515\,\text{g})(4.184\,\text{J/g°C})(2.6°\text{C}) \\ &= 680\,\text{J} \quad \text{(heat absorbed by the water)} \end{aligned}$$

Let x be the metal. Then, since all of the heat absorbed by the water came from the hot metal, we can say that

$$\begin{aligned} q_w &= q_x = (m_x)(sp\,ht_x)\Delta t_x \\ \Delta t_x &= 99.0°\text{C} - 20.6°\text{C} = 78.4°\text{C} \\ 680\,\text{J} &= (68.3820\,\text{g})(sp\,ht_x)(78.4°\text{C}) \\ sp\,ht_x &= 0.13\,\text{J/g°C} \end{aligned}$$

Refer to Table 5.1 and determine that the unknown metal is lead or gold.

PROCEDURE: TRIAL 1

Wear protective glasses.

No waste generated by this experiment.

1. Weigh a dry metal sample and record the mass and the name of the metal on the report form.

2. Carefully slide the sample into a large dry test tube and put a thermometer beside it in the test tube.

3. Attach the test tube to a ring stand and place it into an empty 400 mL beaker as shown in Figure 5.1. Be sure the height of the beaker is adjusted so the hottest part of the burner flame will be on the bottom of the beaker. Do not heat the dry beaker while making this adjustment. The bottom of the test tube should be at least one-half inch above the bottom of the beaker.

4. Fill the beaker with tap water so the height of the water in the beaker is about two inches higher than the top of the metal sample. There should be no water inside the test tube.

5. Begin heating the water in the beaker and continue with the next step(s). As you are working, check the water and note when it starts to boil. Turn down the burner but keep the water gently boiling. Do not do step 11 until the water has been boiling for about ten minutes and the temperature in the test tube has stabilized.

6. Nest two dry styrofoam cups together, weigh them, and record the mass on your report form.

7. Take another metal sample similar to the one you are heating. It does not matter if it is not the same metal. It is a "stand in" for the metal sample you are heating and will not be used during the experiment. Put this "stand in" into the styrofoam cup and add enough distilled water to cover the metal by no more than one-half inch.

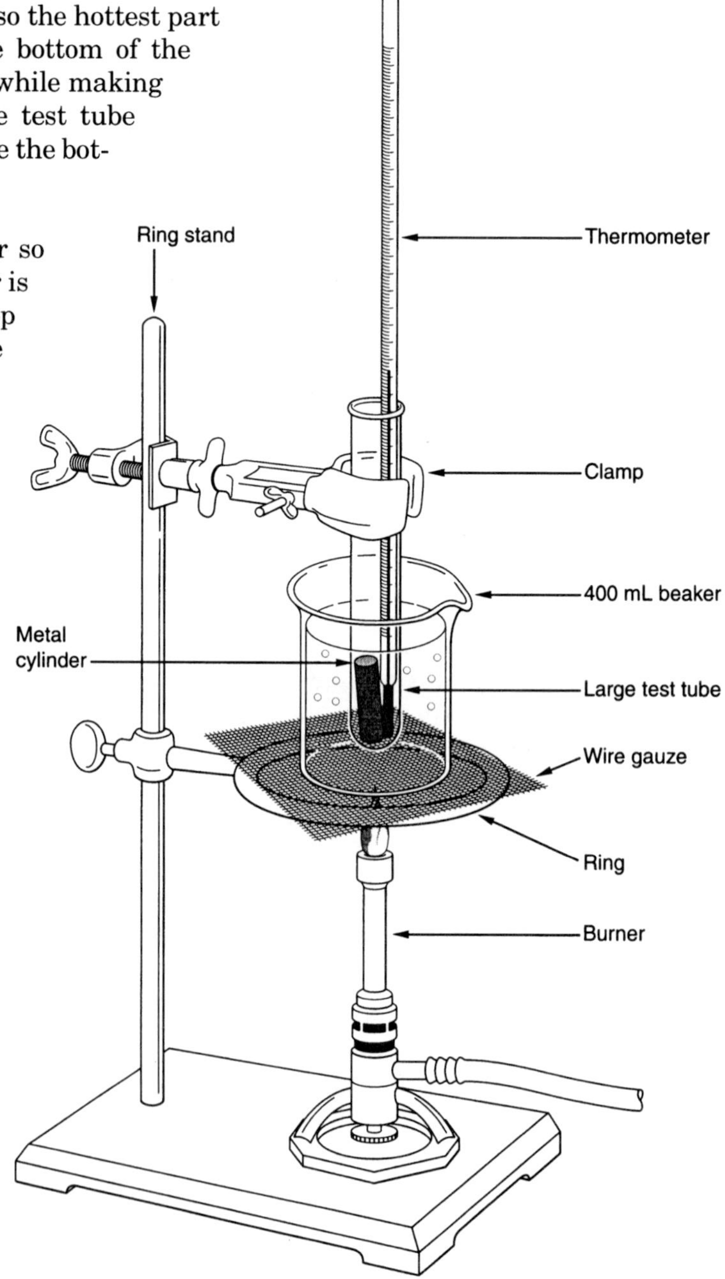

Figure 5.1 The boiler

8. Remove the "stand in" and weigh and record the mass of the styrofoam cups plus the water that you added.

9. Take a cardboard cover for the styrofoam cup and insert a thermometer through the hole. The nested cups with the cardboard cover and thermometer are referred to as a calorimeter, Figure 5.2. If you just leave the calorimeter on the benchtop it might fall over and break the thermometer so put the whole setup into a small beaker to stabilize it. The cardboard cover must rest directly on top of the styrofoam cup and not on the beaker.

10. Measure and record the temperature of the water in the styrofoam cup. Leave the thermometer in the cover until you are ready to transfer the hot metal into the calorimeter.

11. After the water in the beaker has been boiling for 10 minutes and the temperature inside the test tube with the metal has been stable for 5 minutes record the temperature on your report form. Remove the thermometer from the test tube and set it aside so it does not get mixed up with the thermometer used in the calorimeter.

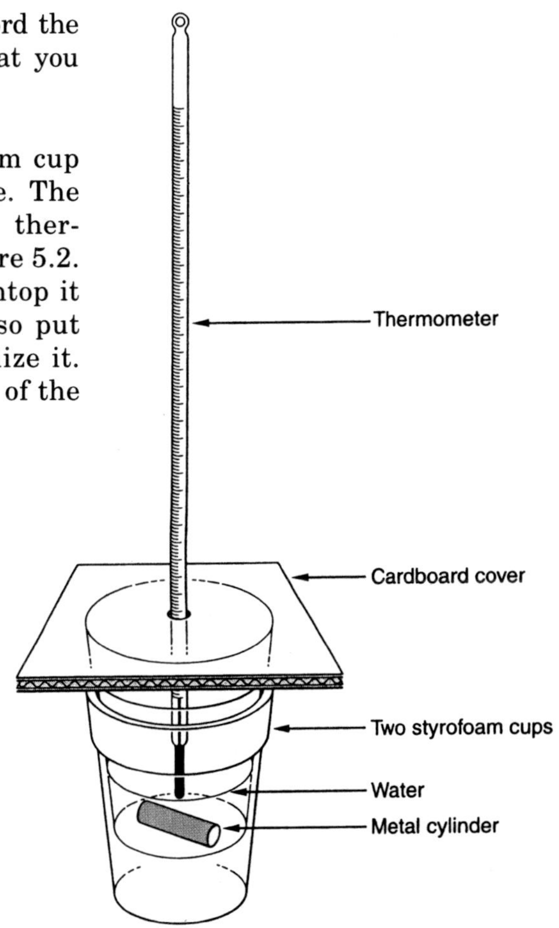

Figure 5.2 The calorimeter (see #9)

12. Now, you are going to transfer the metal from the test tube to the water in the calorimeter. It is important that the transfer take place quickly and carefully to minimize heat loss to the surroundings and to avoid splashing. Remove the cardboard cover and thermometer from the calorimeter. Loosen the clamp on the boiler ring stand, lift the clamp and test tube out of the boiler, and quickly slide the metal into the water in the calorimeter.

13. Immediately, put the cardboard cover with the thermometer back on the styrofoam cup. Stir gently for 2–3 minutes while monitoring the temperature. Record the temperature after it has remained constant for about one minute.

14. Unless instructed otherwise, repeat the experiment with Trial 2.

PROCEDURE: TRIAL 2

15. Repeat this experiment using the same metal sample, but this time use colder distilled water (5–10°C). Do not record the initial temperature of the cold water in the calorimeter until immediately before you add the hot metal. Fill out the report form in the column labeled Trial 2.

16. Look up and record the theoretical value for the specific heat for your metal sample. This can be found in Table 5.1 or in the *Handbook of Chemistry and Physics* if it is a metal not listed in the table.

NAME ____________________

SECTION __________ DATE __________

REPORT FOR EXPERIMENT 5

INSTRUCTOR ____________________

Calorimetry and Specific Heat

Measurements and Calculations

	TRIAL 1	*TRIAL 2*
1. Mass of metal sample	______	______
2. Mass of calorimeter (styrofoam cups)	______	______
3. Mass of styrofoam cups + water	______	______
4. Mass of water (show calculation setup)	______	______
5. Initial water temperature	______	______
6. Temperature of heated metal sample	______	______
7. Final temperature of water and metal	______	______
8. Change in temperature, Δt_w, of the water in the calorimeter (show calculation setup)	______	______
9. Change in temperature, Δt_x, of the metal sample (show calculation setup)	______	______
10. Specific heat ($sp\ ht_w$) of water	*4.184 J/g°C*	*4.184 J/g°C*
11. Heat (q_w) gained by water (show calculation setup)	______	______
12. Heat (q_x) lost by metal sample	______	______
13. Specific heat ($sp\ ht_x$) of the metal (experimental value) (show calculation setup)	______	______

14. Name of metal __________ Theoretical Specific Heat __________

QUESTIONS AND PROBLEMS

1. Why is it important for there to be enough water in the calorimeter to completely cover the metal sample?

2. Why did we heat the metal in a dry test tube rather than in the boiling water?

3. The water in the beaker gets its heat energy from the ______________ and the water in the calorimeter gets its heat energy from the ________________.

4. What is the specific heat in J/g°C for a metal sample with a mass of 95.6 g which absorbs 841 J of energy when its temperature increases from 30.0°C to 98.0°C?

5. What effect does the initial temperature of the water have on the change in temperature of the water after the hot metal is added? Explain your answer.

> Results of scientific experiments must be reproducible when repeated or they do not mean anything. When results are repeated, the experiment is said to have good **precision.** When the results agree with a theoretical value they are described as **accurate.**

6. Which is better, the precision or the accuracy of the experimental specific heats determined for your metal sample? Support your answer with your data.

> Use the data presented in Table 5.1 to answer questions 7–9 and complete the graph on the next page to show the relationship between the atomic mass and the specific heat of the seven metals listed. Make the graph following the guidelines provided in Study Aid 3.

7. a. What is the independent variable? ________________________

   b. What is the dependent variable? ________________________

8. In Table 5.1, what is the range of values for atomic mass? ________________________

9. In Table 5.1, what is the range of values for specific heat?__________

10. Plot the data in Table 5.1 on the graph below. Be sure to include the following:

    a. A title

    b. Placement of the independent and dependent variables on the appropriate axes

    c. Increments for each axis

    d. Labels for each axis

    e. Plotting the data points

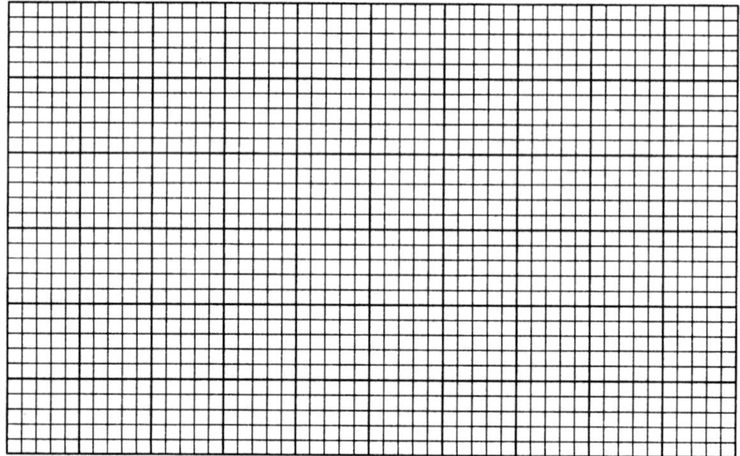

11. Use your graph to summarize the relationship between the atomic mass of metal atoms and the specific heat of a metal.

12. The specific heat was measured for two unknown metal samples. The first sample tested had a specific heat of 0.54 J/g°C. Use your graph to estimate the atomic mass of the metal.

    ____

    The second metal had a specific heat of 0.24 J/g°C. Use your graph again and estimate the atomic mass of this metal.

    ____

EXPERIMENT 6

Freezing Points—Graphing of Data

MATERIALS AND EQUIPMENT

Solids: benzoic acid (C_6H_5COOH) and crushed ice. **Liquid:** glacial acetic acid ($HC_2H_3O_2$). Thermometer, watch or clock with second hand, slotted corks or stoppers.

DISCUSSION

All pure substances, elements and compounds, possess unique physical and chemical properties. Just as one human being can be distinguished from all others by certain characteristics—fingerprints or DNA, for example—it is also possible, through knowledge of its properties, to distinguish any given compound from among the many millions that are known.

A. Melting and Freezing Points of Pure Substances

The melting point and the boiling point are easily determined physical properties that are very useful in identifying a substance. Consequently, these properties are almost always recorded when a compound is described in the chemical literature (textbooks, handbooks, journal articles, etc.). The freezing and melting of a pure substance occurs at the same temperature, measured when the liquid and solid phases of the substance are in equilibrium. When energy is being removed from a liquid in equilibrium with its solid, the process is called freezing; when energy is being added to a solid in equilibrium with its liquid, the process is called melting.

$$\text{liquid} \underset{\substack{+\text{ energy}\\ \text{(melting)}}}{\overset{\substack{\text{(freezing)}\\ -\text{ energy}}}{\rightleftarrows}} \text{solid}$$

In this experiment, we will determine the freezing point of a pure organic compound, glacial acetic acid ($HC_2H_3O_2$). When the experimental freezing point has been determined, it will be compared with the melting point temperature listed in the *Handbook of Chemistry and Physics.*

When heat is removed from a liquid, the liquid particles lose kinetic energy and move more slowly causing the temperature of the liquid to decrease. Finally enough heat is removed and the particles move so slowly that the liquid becomes a solid, often a crystalline solid. The temperature when this happens (the freezing point) is different for different substances.

The amount of energy removed from a quantity of liquid to freeze it, is equal to the amount of energy added to the same quantity of its solid to melt it. Thus, depending on the direction of energy flow, this equilibrium temperature is called the melting point or the freezing point.

B. Freezing Point of Impure Substances

When a substance (solvent) is uniformly mixed with a small amount of another substance (solute), the freezing point of the resulting solution (an "impure substance") will be lower than that of the pure solvent. For example, the accepted freezing point for pure water is 0.0°C. Solutions of salt in water may freeze at temperatures as low as −21°C depending on the amount of salt added to the water. Antifreeze is added to the water in a car radiator to lower the freezing point of the water.

Melting point/freezing point data are of great value in determining the identity and/or purity of substances, especially in the field of organic chemistry. If a sample of a compound melts or freezes appreciably below the known melting point of the pure substance, we know that the sample contains impurities which have lowered the melting point. If the melting point of an unknown compound agrees with that of a known compound, the identity can often be confirmed by mixing the unknown compound with the known and determining the melting point of the mixture. If the melting point of the mixture is the same as that of the known compound, the compounds are identical. On the other hand, a lower melting point for the mixture indicates that the two compounds are not identical.

C. Supercooling During Freezing

Frequently when a substance is being cooled, the temperature will fall below the true freezing point before crystals begin to form. This phenomenon is known as supercooling because the substance is cooled below its freezing point without forming a solid. Supercooling is more likely to occur if the liquid remains very still and undisturbed as its temperature is lowered. When the system is disturbed in any way, for example, by stirring or jarring, crystallization occurs rapidly throughout the system. As the crystals form, heat is released (called the heat of crystallization) and the temperature rises quickly to the freezing point of the substance. Thus, supercooling does not change the freezing point of the substance.

D. Freezing Point Determinations

You will do three freezing point determinations during this experiment using the setup in Figure 6.1.

Trial 1. Freezing point determination of pure glacial acetic acid WITH STIRRING. This will usually eliminate supercooling.

Trial 2. Freezing point determination of pure glacial acetic acid WITHOUT STIRRING. This should enhance the possibility of supercooling but does not guarantee it.

Trial 3. Freezing point determination of acetic acid (the solvent) after benzoic acid (a solute) has been dissolved in it. This will be done WITHOUT STIRRING to enhance supercooling again.

The time/temperature data will be graphed and the freezing point for each trial read from the graph.

PROCEDURE

Wear protective glasses.

> **NOTES:** Since water and other contaminants will influence the freezing points in this experiment, use only clean, dry equipment.
>
> Read and record all temperatures to the nearest 0.1°C.

A. Freezing Point Determination of Pure Glacial Acetic Acid

Trial 1: With stirring

1. Fasten a utility clamp to the top of a clean, dry test tube. Position this clamp-tube assembly on a ring stand so that the bottom of the tube is about 20 cm above the ring stand base.

2. Obtain a slotted one-hole cork (or stopper) to fit the test tube (see Figure 6.1). Insert a thermometer in the cork and position it in the test tube so that the end of the bulb is about 1.5 cm from the bottom of the test tube. Turn the thermometer so that the temperature scale can be read in the slot.

3. Take your test tube, the cork/thermometer and a graduated cylinder to the fume hood. Measure out 10. mL of glacial acetic acid. Pour it into the test tube and close the test tube with the cork/thermometer. Glacial acetic acid is irritating and harmful if inhaled so keep the test tube stoppered while you work outside the hood at your bench. Rinse the graduated cylinder with water immediately.

4. Reclamp the test tube to your ring stand to minimize the risk of spilling. Make sure the thermometer bulb is covered by the acid and adjust the temperature of the acetic acid to approximately 25°C by warming or cooling the tube in a beaker of water.

5. Fill a 400 mL beaker about three-quarters full of crushed ice; add cold water until the ice is almost covered. Position the beaker of ice and water on the ring stand base under the clamped tube-thermometer assembly.

6. Read the temperature of the acetic acid and record as the 0.0 minute time reading in the Data Table. Now loosen the clamp on the ring stand and observe the second hand of your watch or clock. As the second hand crosses 12, lower the clamped tube-thermometer assembly so that all of the acetic acid in the tube is below the surface of the ice water. Fasten the clamp to hold the tube in this position.

7. Loosen the cork on the tube and stir (during Trial 1 only) the acid with the thermometer, keeping the bulb of the thermometer completely immersed in the acid. Take accurate temperature readings at 30-second intervals as the acid cools. (Zero time was when the second hand crossed 12.) Stop stirring and center the thermometer bulb in the tube as soon as you are sure that crystals are forming in the acid (one to four minutes). Circle the temperature reading when the crystals were first observed.

8. Continue to take temperature readings at 30-second intervals until a total time of 12 minutes has elapsed or until the entire volume of liquid becomes solidified. After that occurs, read the temperature for an additional 2 minutes (4 time intervals) and continue with the next step.

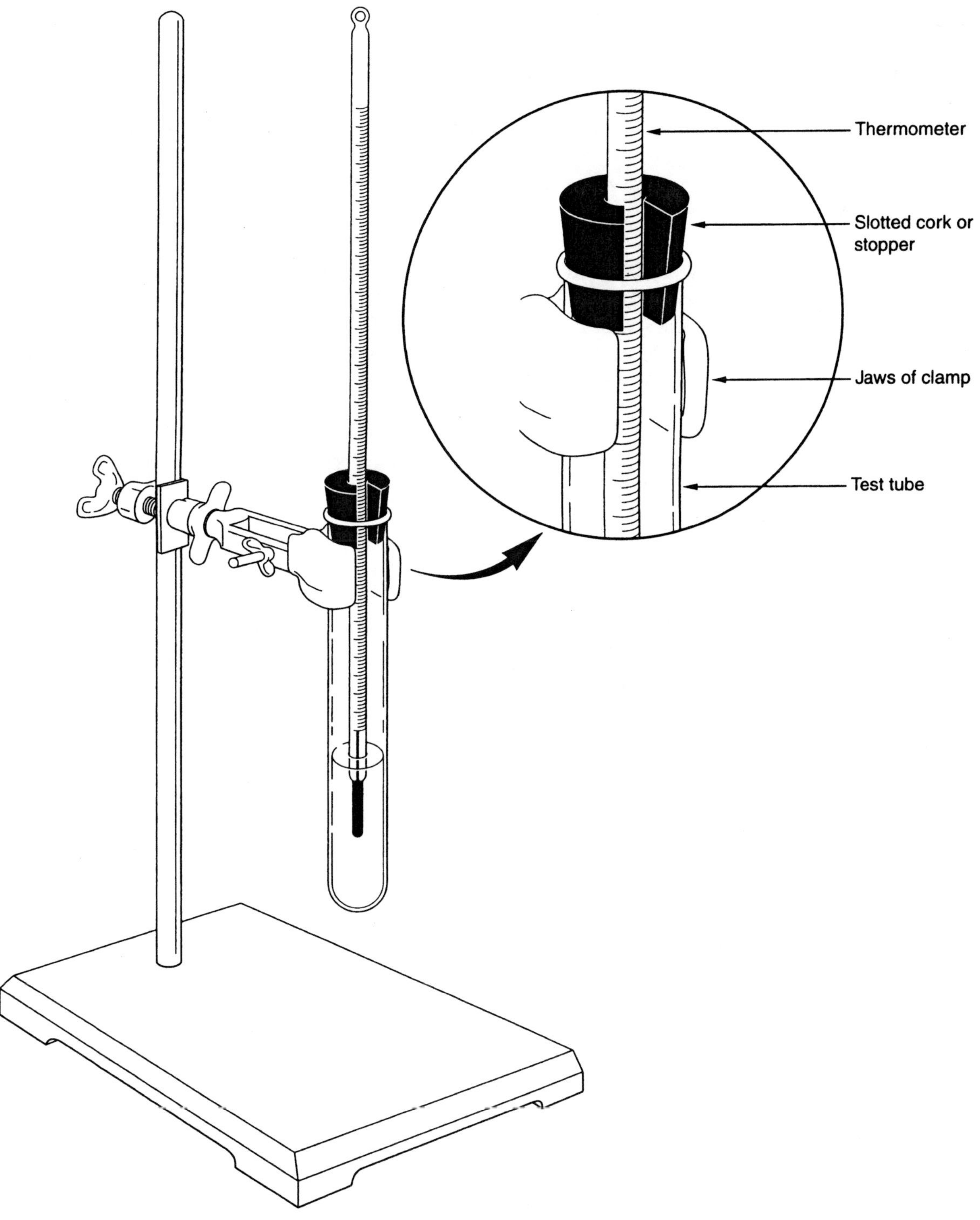

Figure 6.1 Setup for freezing-point determination

9. After completing the temperature readings, remove the test tube-thermometer assembly from the ice bath, keeping the thermometer in place. Immerse the lower portion of the test tube in a beaker of warm water to melt the frozen acetic acid. Do not discard this acid; it will be used in Trials 2 and 3.

Trial 2: Without stirring

10. Repeat steps 4–9 with the following changes:

    a. Replenish the ice bath as in step 5.

    b. After submerging the tube in the ice bath, do NOT stir. Do NOT touch or move the apparatus in any way

    c. If the temperature goes down to about 4°C or lower without the formation of acetic acid crystals and remains there, touch the thermometer and move it until crystals form which usually happens quickly. When you do this be very observant of the temperature changes. Continue to record temperature readings for the full 12 minutes or until the temperature stabilizes after crystallization for 5 minutes.

B. Freezing Point Determination of An Acetic Acid/Benzoic Acid Solution

Trial 3: Without stirring

11. Weigh approximately 0.50 g (between 0.48 and 0.52 g) of benzoic acid crystals. Now remove the thermometer from the test tube of acetic acid and lay it on the table) being careful not to contaminate the thermometer or lose any acid. Carefully add all of the benzoic acid to the acetic acid. Stir gently with the thermometer until all of the crystals have dissolved. Stir for an additional minute or two to ensure a uniform solution. Adjust the temperature of the solution to approximately 25°C.

12. Repeat step 10.

WASTE DISPOSE OF PROPERLY

13. Warm the test tube to change the solid to a liquid and dispose of the acetic acid/benzoic acid solution in the waste container provided. Rinse the test tube with water and pour the liquid down the sink.

C. Graphing Temperature Data

Graph the three sets of data using the graph paper in the report form or prepare a computer graph. If necessary, review the instructions for preparing a graph in Study Aid 3.

180°C

<180°C

melting points

Ethylene Glycol - decreases freezing pt. of H_2O

NAME ____________________

SECTION __________ DATE __________

REPORT FOR EXPERIMENT 6

INSTRUCTOR ____________________

Freezing Points–Graphing of Data

Data Table

	Pure Acetic Acid	**Pure Acetic Acid**	**Impure Acetic Acid**
time, minutes	**temp, °C** WITH STIRRING	**temp, °C** WITHOUT STIRRING	**temp, °C** WITHOUT STIRRING
0.0			
0.5			
1.0			
1.5			
2.0			
2.5			
3.0			
3.5			
4.0			
4.5			
5.0			
5.5			
6.0			
6.5			
7.0			
7.5			
8.0			
8.5			
9.0			
9.5			
10.0			
10.5			
11.0			
11.5			
12.0			

Graphing of Freezing Point Data

Plot your data on the graph paper or the computer using a legend as follows:

△ = Pure acetic acid with stirring

▲ = Pure acetic acid without stirring

○ = Acetic acid/benzoic acid solution without stirring

Draw rectangles around the portions of your curves that show supercooling.

 NAME ____________

QUESTIONS

Use your graph to answer the questions 1–3.

1. a. At what temperature did crystals first form in Trial 1? ________

   b. Where did the temperature stabilize after supercooling in Trial 2? ________

   c. What is your experimental freezing point of glacial acetic acid? ________

   d. What is the theoretical freezing point of glacial acetic acid? ________
   (Consult the *Handbook of Chemistry and Physics*)

2. How many degrees was the freezing point depressed by the benzoic acid? ________

   Do this by estimating to the nearest 0.1 degree the number of degrees between the flattest (most nearly horizontal) portions of the curves. Mark the area on the graph with an arrow (↓) to show where this temperature difference estimate was made.

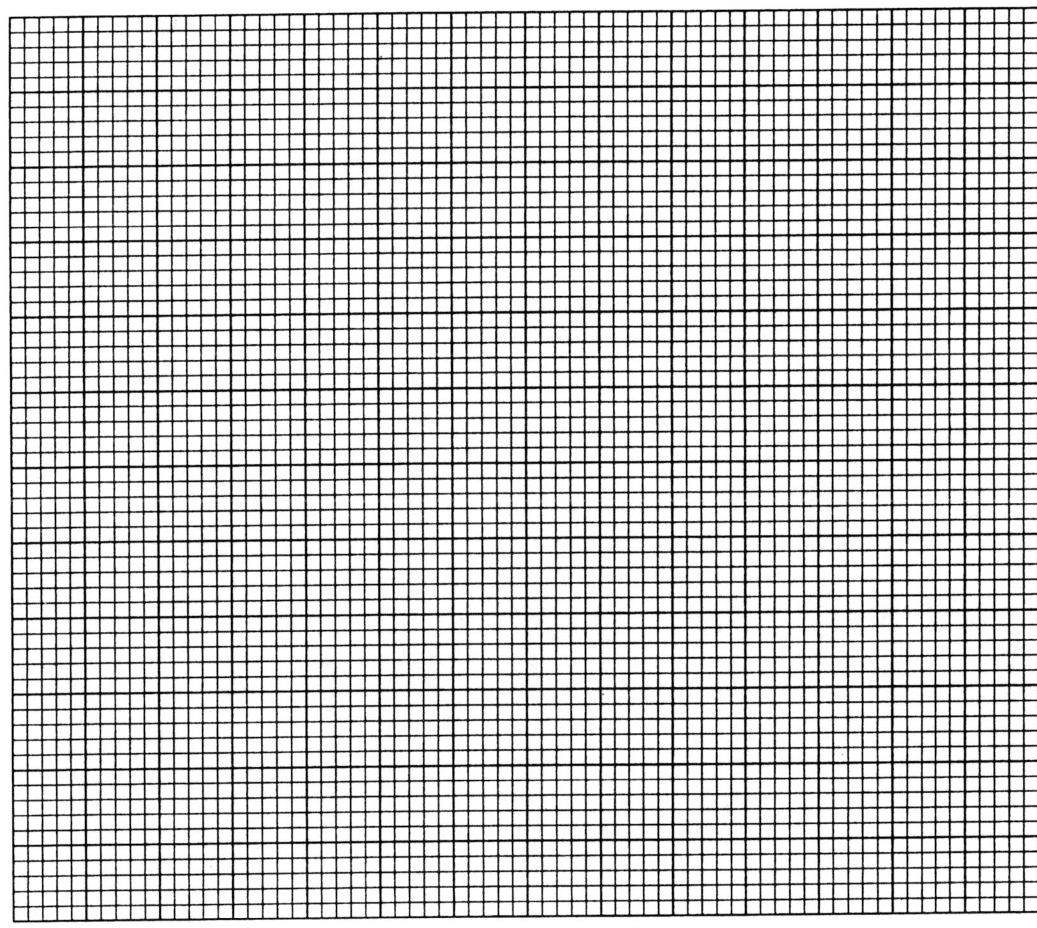

3. a. What is the effect of stirring on the freezing point of pure acetic acid?

   b. What is the effect of stirring on supercooling?

4. a. What do the melting point and freezing point of a substance have in common?

   b. What is the difference between the melting and freezing of a substance?

5. When the solid and liquid phases are in equilibrium, which phase, solid or liquid contains the greater amount of energy? Explain the rationale for your answer.

EXPERIMENT 7

Water in Hydrates

MATERIALS AND EQUIPMENT

Solids: finely ground copper(II) sulfate pentahydrate ($CuSO_4 \cdot 5H_2O$), and unknown hydrate. Cobalt chloride test paper, clay triangle, crucible and cover, 25 × 200 mm ignition test tube, watch glass.

DISCUSSION

Many salts form compounds in which a definite number of moles of water are combined with each mole of the anhydrous salt. Such compounds are called **hydrates.** The water which is chemically combined in a hydrate is referred to as **water of crystallization** or **water of hydration.** The following are representative examples:

$$CaSO_4 \cdot 2H_2O, \quad CoCl_2 \cdot 6H_2O, \quad MgSO_4 \cdot 7H_2O, \quad Na_2CO_3 \cdot 10H_2O$$

In a hydrate the water molecules are distinct parts of the compound but are joined to it by bonds that are weaker than either those forming the anhydrous salt or those forming the water molecules. In the formula of a hydrate a dot is commonly used to separate the formula of the anhydrous salt from the number of molecules of water of crystallization. For example, the formula of calcium sulfate dihydrate is written $CaSO_4 \cdot 2H_2O$ rather than $CaSO_6H_4$.

Hydrated salts can usually be converted to the anhydrous form by careful heating:

$$\text{Hydrated Salt} \xrightarrow{\Delta} \text{Anhydrous salt} + \text{water}$$

Hydrated salts can be studied qualitatively and quantitatively. In the **qualitative** part of this experiment we will observe some of properties of the liquid (water) driven off by heating the sample. In the **quantitative** part of the experiment we will determine **how much** water was in the hydrate by measuring the amount of water driven off by heating.

To make certain that all of the water in the original sample has been driven off, chemists use a technique known as **heating to constant weight.** Since time expended for this is limited, constant weight is essentially achieved when the sample is heated and weighed in successive heatings until the weight differs by no more than 0.05 g. Thus, if the second weighing is no more than 0.05 g less than the first heating, a third heating is not necessary because the sample has been heated to constant weight (almost). This is a very good reason to follow directions meticulously when heating. If the sample is not heated long enough or at the correct temperature, all of the water may not be driven off completely in the first heating.

Hence it is possible to determine the percentage of water in a hydrated salt by determining the amount of mass lost (water driven off) when a known mass of the hydrate is heated to constant weight.

$$\text{Percentage water} = \left(\frac{\text{Mass lost}}{\text{Mass of sample}}\right)(100)$$

It is possible to condense the vapor driven off the hydrate and demonstrate that it is water by testing it with anhydrous cobalt(II) chloride ($CoCl_2$). Anhydrous cobalt(II) chloride is blue but reacts with water to form the red hexahydrate, $CoCl_2 \cdot 6H_2O$.

PROCEDURE

Wear protective glasses.

A. Qualitative Determination of Water

1. Fold a 2.5 × 20 cm strip of paper lengthwise to form a V-shaped trough or chute. Load about 4 g of finely ground copper(II) sulfate pentahydrate in this trough, spreading it evenly along the length of the trough.

2. Clamp a **dry** 25 × 200 mm ignition test tube so that its mouth is 15–20 degrees **above the horizontal** (Figure 7.1a). Insert the loaded trough into the tube. Rotate the tube to a nearly vertical position (Figure 7.1b) to deposit the copper(II) sulfate in the bottom of the tube. Tap the paper chute gently if necessary, but make sure that no copper sulfate is spilled and adhering to the sides of the upper part of the tube.

3. Remove the chute and turn the tube until it slants mouth downward at an angle of 15–20 degrees **below the horizontal** (Figure 7.1c). Make sure that all of the copper(II) sulfate remains at the bottom of the tube. To obtain a sample of the liquid that will condense in the cooler part of the tube, place a clean, dry test tube, held in an upright position in either a rack or an Erlenmeyer flask, just below the mouth of the tube containing the hydrate.

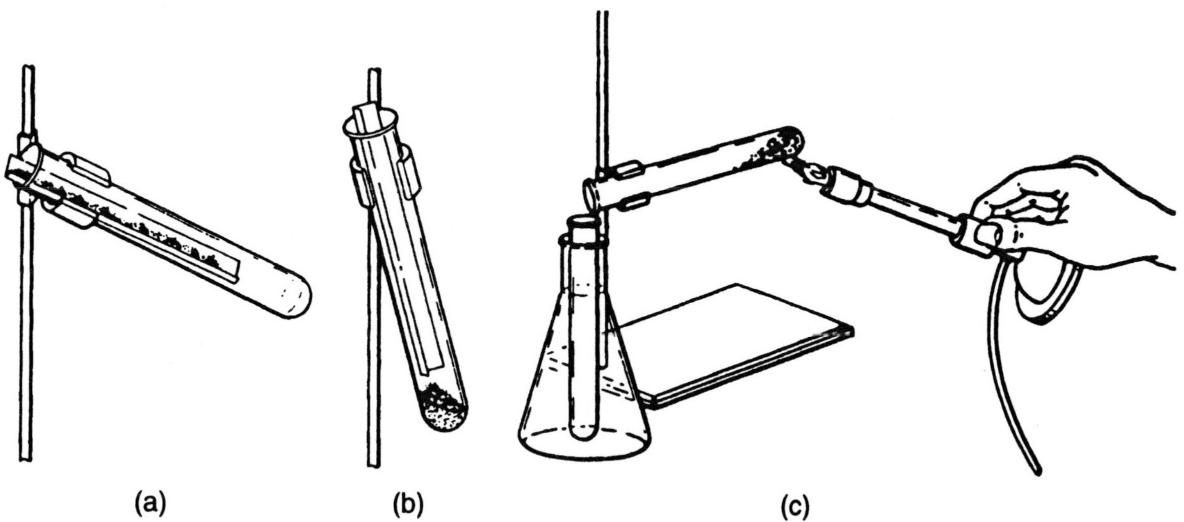

Figure 7.1 Setup for dehydration of a hydrate

4. Heat the hydrate gently at first to avoid excessive spattering. Gradually increase the rate of heating, noting any changes that occur and collecting some of the liquid that condenses in the cooler part of the tube. Continue heating until the blue color of the hydrate has disappeared, but do not heat until the residue in the tube has turned black. Finally warm the tube over its entire length—without directly applying the flame to the clamp—for a minute

or two to drive off most of the liquid that has condensed on the inner wall of the tube. Allow the tube and contents to cool.

NOTE: At excessively high temperatures (above 600°C) copper(II) sulfate decomposes; sulfur trioxide is driven off and the black copper(II) oxide remains as a residue.

Observe and record the appearance and odor of the liquid that has been collected.

5. While the tube is cooling, dry a piece of cobalt chloride test paper by holding it with tongs about 20 to 25 cm above a burner flame; that is, close enough to heat but not close enough to char or ignite the paper. When properly dried, the test paper should be blue. Using a clean stirring rod, place a drop of the liquid collected from the hydrate on the dried cobalt chloride test paper. For comparison place a drop of distilled water on the cobalt chloride paper. Record your observations.

6. Empty the anhydrous salt residue in the tube onto a watch glass and divide it into two portions. Add 3 or 4 drops of the liquid collected from the hydrate to one portion and 3 or 4 drops of distilled water to the other. Compare and record the results of these tests.

Dispose of solid residues in the waste heavy metal container provided.

B. Quantitative Determination of Water in a Hydrate

NOTES:

1. **Weigh crucible and contents to the highest precision with the balance available to you.**
2. Since there is some inaccuracy in any balance, use the same balance for successive weighings of the same sample. When subtractions are made to give mass of sample and mass lost, the inaccuracy due to the balance should cancel out.
3. Handle crucibles and covers with tongs only, after initial heating.
4. Be sure crucibles are at or near room temperature when weighed.
5. **Record all data directly on the report form as soon as you obtain them.**

1. Obtain a sample of an unknown hydrate, as directed by your instructor. Be sure to record the identifying number.

2. Weigh a clean, dry crucible and cover to the highest precision of the balance.

3. Place between 2 and 3 g of the unknown into the weighed crucible. Cover and weigh the crucible and contents.

4. Place the covered crucible on a clay triangle; adjust the cover so that is slightly ajar, to allow the water vapor to escape (see Figure 7.2); and **very gently** heat the crucible for about 5 minutes. Readjust the flame so that a sharp, inner-blue cone is formed. Heat for another 12 minutes with the tip of the inner-blue cone just touching the bottom of the crucible. The crucible bottom should become dull red during this period.

5. After this first heating is completed, close the cover, cool (about 10 minutes), and weigh.

6. To determine if all the water in the sample was removed during the initial heating, reheat the covered crucible and contents for an additional 6 minutes at maximum temperature; cool and reweigh. If the sample was heated to constant weight the results of the last two weighings should agree within 0.05 g. If the decrease in mass between the two weighings is greater than 0.05 g, repeat the heating and weighing until the results of two successive weighings agree to within 0.05 g.

7. Calculate the percentage of water in your sample on the basis of the *final* weighing.

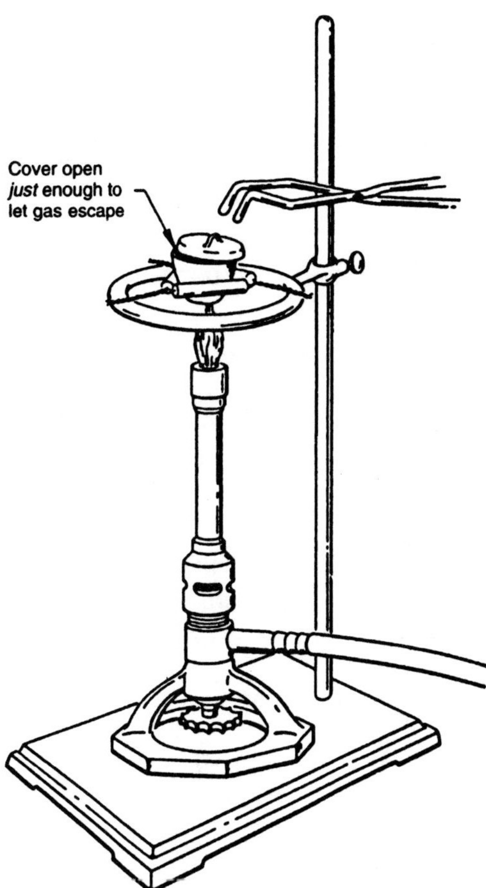

Figure 7.2 Method of heating a crucible

Dispose of the solid residue in the waste heavy metal container provided. Return the unused portion of your unknown to the instructor.

NAME Ellen Jaggers

SECTION ________ DATE 10-16-14

REPORT FOR EXPERIMENT 7

INSTRUCTOR Prof. Jaggi

Water in Hydrates

A. Qualitative Determination of Water

1. Describe the appearance and odor of the liquid obtained by heating copper(II) sulfate pentahydrate. clear, odorless liquid.

2. Compare the results observed when testing the liquid from the hydrate and distilled water with the cobalt chloride paper and the anhydrous salt by completing the table below.

Property Observed	Cobalt Chloride Paper	Anhydrous $CuSO_4$
Color before adding liquid(s) to	blue	white
Color after adding distilled water to	whitish pink	sky blue
Color after adding liquid from hydrate to	whitish pink	sky blue
Temperature change after adding distilled water	N/A	N/A
Temperature change after adding liquid from hydrate	N/A	N/A

B. Quantitative Determination of Water in a Hydrate

1. Mass of crucible and cover 42.48g
2. Mass of crucible, cover, and sample 44.61g
3. Mass of crucible, cover, and sample after 1st heating 43.72g
4. Mass of crucible, cover, and sample after 2nd heating 43.72g
5. Mass of crucible, cover, and sample after 3rd heating (if needed) ________
6. Mass of original sample 2.13g
   Show calculation setup: 44.61 − 42.48
7. Total mass lost by sample during heating 0.89g
   Show calculation setup: 44.61 − 43.72
8. Percentage water in sample — Sample No. 15 — 42%
   Show calculation setup:

$.89 = \frac{x}{100}(2.13)$

$\left(\frac{100}{2.13}\right).89 = x$

41.784 = x round to 2 sig figs x = 42%

 NAME Ellen Jaggers

QUESTIONS AND PROBLEMS

1. What evidence did you see that indicated the liquid obtained from the copper (II) sulfate pentahydrate was water?
   (a) It was a clear odorless liquid when observed.

   (b) It turned the cobalt chloride paper the same color as did the distilled water.

2. What was the evidence of a chemical reaction when the anhydrous salt samples were treated with the liquid obtained from the hydrate and with water?
   (a) The color changed from white to sky blue.

   (b)

3. Write a balanced chemical equation for the decomposition of copper(II) sulfate pentahydrate.
   $CuSO_4 \cdot H_2O \xrightarrow{\Delta} CuSO_4 + H_2O$

4. When the unknown was heated, could the decrease in mass have been partly due to the loss of some substance other than water? Explain.
   No. The unknown is a hydrate, which means it consists of a salt and water. The water is separated by heating, because it evaporates. No other components of the hydrate can evaporate from the crucible.

5. A student heated a hydrated salt sample with an initial mass of 4.8702 g, After the first heating, the mass had decreased to 3.0662 g.
   (a) If the sample was heated to constant weight after reheating, what is the minimum mass that the sample can have after the second weighing? Show how you determined your answer.
   3.0662 − .05 3.0162g

   (b) The student determined that the mass lost by the sample was 1.8053. What was the percent water in the original hydrated sample? Show calculation setup.
   $1.8053 = \frac{x}{100}(4.8702)$ 37.068%

   $\left(\frac{100}{4.8702}\right) 1.8053 = x$

   $37.06829 = x$ round to 5 sig figs = 37.068%

EXPERIMENT 9

Properties of Solutions

MATERIALS AND EQUIPMENT

Solids: ammonium chloride (NH_4Cl), barium chloride ($BaCl_2$), barium sulfate ($BaSO_4$), fine and coarse crystals of sodium chloride (NaCl), and sodium sulfate (Na_2SO_4). **Liquids:** decane ($C_{10}H_{22}$), isopropyl alcohol (C_3H_7OH), and kerosene. **Solutions:** saturated iodine-water (I_2), and saturated potassium chloride (KCl).

DISCUSSION

Solute, Solvent, and Solution

The term **solution** is used in chemistry to describe a homogeneous mixture in which at least one substance (the **solute**) is dissolved in another substance (the **solvent**). The solvent is the substance present in greater quantity and the name of the solution is taken from the name of the solute. Thus, when sodium chloride is dissolved in water, sodium chloride is the solute, water is the solvent, and the solution is called a sodium chloride solution.

In this experiment we will be working with two common types of solutions: those in which a solid solute is dissolved in a liquid solvent (water), and a few in which a liquid solute is dissolved in a liquid solvent.

Like other mixtures, a solution has variable composition, since more or less solute can be dissolved in a given quantity of a solvent. The amount of solute that remains uniformly dispersed throughout the solution after mixing is referred to as the **solution concentration** and can be expressed in many different ways. The maximum concentration that a solution can have varies depending on many factors, including the temperature, the kind of particles in the solute, and interactions between the solute particles and the solvent. In general, water, which is polar, is a better solvent for inorganic than for organic substances. On the other hand, nonpolar solvents such as benzene, decane, and ether are good solvents for many organic substances that are practically insoluble in water.

Dissolved solute particles can be either molecules or ions and their size is of the order of 10^{-8} to 10^{-7} cm (1-10 Å). Many substances will react chemically with each other only when they are dissociated into ions in solution. For example, when the two solids sodium chloride (NaCl) and silver nitrate ($AgNO_3$) are mixed, no detectable reaction is observed. However, when aqueous solutions of these salts are mixed, their component ions react immediately to form a white precipitate (AgCl).

The rate at which a solute and solvent will form a solution depends on several factors, all of which are related to the amount of contact between the solute particles and the solvent. A solid can dissolve only at the surface that is in contact with the solvent. Any change which increases that contact will increase the rate of solution and vice versa. Thus, the rate of dissolving a solid solute depends on:

1. The particle size of the solute
2. Agitation or stirring of the solution
3. The temperature of the solution
4. The concentration of the solute in solution

Solubility, Miscibility, and Concentration

The term **solubility** refers to the maximum amount of solute that will dissolve in a specified amount of solvent under stated conditions. At a specific temperature, there is a limit to the amount of solute that will dissolve in a given amount of solvent.

Solubility can be expressed in a relative, qualitative way. For example a solute may be very soluble, moderately soluble, slightly soluble, or insoluble in a given solvent at a given temperature. Table 8.1 shows how temperature effects the amount of four different salts that dissolve in 100 g of water.

Table 9.1
Temperature Effect on Solubility of Four Salts in Water, g solute/100 g water

	0°C	10°C	20°C	30°C	40°C	50°C	60°C	70°C	80°C	90°C	100°C
KCl	27.6	31.0	34.0	37.0	40.0	42.6	45.5	48.3	51.1	54.0	55.6
NaCl	35.7	35.8	36.0	36.3	36.6	37.0	37.3	37.8	38.4	39.0	39.8
KBr	53.5	59.5	65.2	70.6	75.5	80.2	85.5	90.0	95.0	99.2	104.0
$BaCl_2$	31.6	33.3	35.7	38.2	40.7	43.6	46.6	49.4	52.6	55.7	58.8

The term **miscibility** describes the solubility of two liquids in each other. When both the solute and solvent are liquids, their solubility in each other is described as miscible (soluble) or immiscible (insoluble). For example, ethyl alcohol and water are miscible; oil and water are immiscible.

The **concentration** of a solution expresses how much solute is dissolved in solution and can be expressed several ways:

1. **Dilute vs. Concentrated:** a dilute solution contains a relatively small amount of solute in a given volume of solution; a concentrated solution contains a relatively large amount of solute per unit volume of solution.

2. **Saturated vs. Unsaturated vs. Supersaturated:**

a. A **saturated** solution contains as much dissolved solute as possible at a given temperature and pressure. The dissolved solute is in equilibrium with undissolved solute. A saturated solution can be dilute or concentrated. The solutions described in Table 8.1 are saturated at each temperature.

$$\text{Solute(solid)} \rightleftharpoons \text{Solute(dissolved)}$$

b. **Unsaturated** solutions contain less solute per unit volume than the corresponding saturated solution. Thus, more solute will dissolve in an unsaturated solution (until saturation is reached).

c. **Supersaturated** solutions contain more dissolved solute than is normally present in the corresponding saturated solution. However, a supersaturated solution is in a very unstable state and will form a saturated solution if disturbed. For example, when a small crystal of the dissolved salt is dropped into a supersaturated solution, crystallization begins at once and salt precipitates until a saturated solution is formed.

3. **Mass-percent Solution** is a quantitative expression of concentration expressed as the percent by mass of the solute in a solution. For example, a 10% sodium hydroxide solution contains 10 g of NaOH in 100 g of solution (10 g NaOH + 90 g H_2O); 2 g NaOH in 20 g of solution (2 g NaOH + 18 g H_2O). The formula for calculating mass percent is:

$$\text{Mass percent} = \left(\frac{\text{g solute}}{\text{g solute} + \text{g solvent}}\right)(100)$$

4. **Mass per 100 g solvent** is another quantitative expression of concentration (and the one used in Table 8.1). It is not the same as the Mass percent concentration above because the units are g solute/100 g solvent. Thus, for the 10% NaOH solution described in No. 3, the g NaOH/100 g H_2O would be calculated as follows:

$$\left(\frac{10\,\text{g NaOH}}{90\,\text{g}\,H_2O}\right)(100) = \frac{11\,\text{g NaOH}}{100\,\text{g}\,H_2O}$$

5. **Molarity** is the most common quantitative expression of concentration. Molarity is the number of moles (molar mass) of solute per liter of solution. Thus a solution containing 1 mole of NaOH (40.00 g) per liter is 1 molar (abbreviated 1 M). The concentration of a solution containing 0.5 mole in 500 mL (0.5 L) is also 1 M. The formula for calculating molarity is:

$$\text{Molarity} = \frac{\text{moles of solute}}{\text{liter of solution}} = \frac{\text{moles}}{\text{liter}}$$

PROCEDURE

Wear protective glasses.

A. Concentration of a Saturated Solution

Use the same balance for all weighings.
Make all weighings to the highest precision of the balance.

1. Prepare a water bath with a 400 mL beaker half full of tap water and heat to boiling. (See Figure 1.6.)

2. Weigh an empty evaporating dish. Obtain 6 mL of saturated potassium chloride solution and pour it into the dish. Weigh the dish with the solution in it and record these masses on the report form.

3. Place the evaporating dish on the beaker of boiling water and continue to boil until the potassium chloride solution has evaporated almost to dryness (about 25 to 30 minutes), **adding more water to the beaker as needed.**

While the evaporation is proceeding, continue with other parts of the experiment.

4. Remove the evaporating dish and beaker from the wire gauze and dry the bottom of the dish with a towel. Put the dish on the wire gauze and heat gently for 1-2 minutes to evaporate the last traces of water. Do not heat too strongly because at high temperatures there is danger of sample loss by spattering.

5. Allow the dish with dry potassium chloride to cool on the Ceramfab pad for 5 to 10 minutes and weigh. To be sure that all the water has evaporated from the potassium chloride, put the dish back on the wire gauze and heat gently again for 1-2 minutes.

6. Allow the dish to cool again on the Ceramfab pad for 5 to 10 minutes and reweigh. The second weighing should be no more than 0.05 g less than the first weighing. Otherwise a third heating and weighng should be done.

7. Add water to the residue in the dish to redissolve the potassium chloride. Pour the solution into the sink and flush generously with water.

B. Relative Solubility of a Solute in Two Solvents

1. Add about 2 mL of decane and 5 mL of water to a test tube, stopper it, and shake gently for about 5 seconds. Allow the liquid layers to separate and note which liquid has the greatest density.

2. Now, add 5 mL of saturated iodine-water to the test tube, note the color of each layer, insert the stopper, and shake gently for about 20 seconds. Allow the liquids to separate and again note the color of each layer.

3. Dispose of the mixture in this test tube in the bottle labeled **Decane Waste.**

C. Miscibility of Liquids

1. Take three dry test tubes and add liquids to each as follows:

   a. 1 mL kerosene and 1 mL isopropyl alcohol

   b. 1 mL kerosene and 1 mL water

   c. 1 mL water and 1 mL isopropyl alcohol

2. Stopper each tube and mix by shaking for about 5 seconds. Note which pairs are miscible. Dispose of the kerosene mixtures (a and b) in the bottle labeled **Kerosene Waste.** Dispose the contents in test tube (c) in the sink.

D. Effect of Particle Size on Rate of Dissolving

1. Fill a dry test tube to a depth of about 0.5 cm with fine crystals of sodium chloride. Fill another dry tube to the same depth with coarse sodium chloride crystals. Add 10 mL of tap water to each tube and stopper. Shake both tubes at the same time, noting the number of seconds required to dissolve the salt in each tube. (Don't shake the tubes for more than two minutes.)

2. Dispose of these solutions in the sink.

E. Effect of Temperature on Rate of Dissolving

1. Weigh two 0.5 g samples of fine sodium chloride crystals.

2. Take a 100 mL and a 150 mL beaker and add 50 mL tap water to each. Heat the water in the 150 mL beaker to boiling and allow it to cool for about 1 minute.

3. Add the 0.5 g samples of salt to each beaker and observe the time necessary for the crystals to dissolve in the hot water (do not stir).

4. As soon as the crystals are dissolved in the hot water, take the beaker containing the hot solution in your hand, slowly tilt it back and forth, and observe the layer of denser salt solution on the bottom. Repeat with the cold-water solution.

5. Dispose of these solutions in the sink.

F. Solubility versus Temperature; Saturated and Unsaturated Solutions

1. Label four weighing boats or papers as follows and weigh the stated amounts onto each one.

   a. 1.0 g NaCl   b. 1.4 g NaCl   c. 1.0 g NH_4Cl d. 1.4 g NH_4Cl

2. Record observations in the table provided on the report form as you proceed through 3-6.

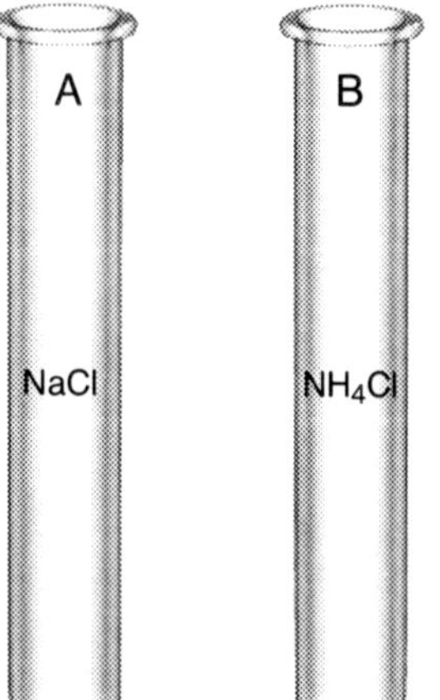

3. Add the 1.0 g samples of NaCl and NH_4Cl to separate tubes labeled A and B as shown. Add 5 mL of distilled water to each, stopper and shake until each salt is dissolved.

4. Now add 1.4 g NaCl to test tube A. Add 1.4 g NH_4Cl to test tube B. Stopper and shake for about 3 minutes. Note whether all of the crystals have dissolved.

5. Place both tubes (unstoppered) into a beaker of boiling water, shake occasionally, and note the results after about 5 minutes.

white powder and white solid on bottom

6. Remove the tubes and cool in running tap water for about 1 minute. Let stand for a few minutes and record what you observe.

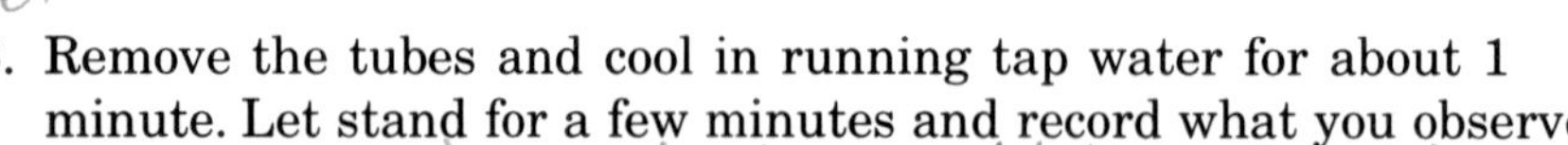

7. Dispose of these solutions in the sink. Flush generously with water.

G. Ionic Reactions in Solution

1. Into four labeled test tubes, place pea-sized quantities of the following salts, one salt in each tube: (a) barium chloride, (b) sodium sulfate, (c) sodium chloride, (d) barium sulfate.

2. Add 5 mL of water to each tube, stopper, and shake to dissolve. One of the four salts does not dissolve.

3. Mix the barium chloride and sodium sulfate solutions together. Note the results. (Sodium chloride and barium sulfate are the products of this reaction.)

Dispose of all tubes containing barium in the waste bottle provided. The remaining tubes can be rinsed in the sink.

NAME Ellen Jaggers

SECTION ________ DATE ____________

REPORT FOR EXPERIMENT 9

INSTRUCTOR ____________________

Properties of Solutions

A. Concentration of Saturated Solution

1. Mass of empty evaporating dish 44.08 g
2. Mass of dish + saturated potassium chloride solution 50.85 g
3. Mass of dish + dry potassium chloride, 1st heating 47.71 g
4. Mass of dish + dry potassium chloride, 2nd heating ________
5. Mass of saturated potassium chloride solution 6.77 g

   Show Calculation Setup

   50.85 - 44.08

6. Mass of potassium chloride in the saturated solution 3.63 g

   Show Calculation Setup

   6.77 - 3.14

7. Mass of water in the saturated potassium chloride solution 3.14 g

   Show Calculation Setup

   50.85 - 47.71

8. Mass percent of potassium chloride in the saturated solution 53.6%

   Show Calculation Setup

   $\frac{3.63}{6.77} \times 100$

   $\frac{KCl}{Water}$

9. Grams of potassium chloride per 100 g of water (experimental) in the original solution. 53.6 g

   Show Calculation Setup

   $\frac{KCl\ g}{100\ g} = \frac{3.63}{6.77}$

10. Grams of potassium chloride per 100 g of water (theoretical) (From Table 8.1) at 20°C. 34.0 g

B. Relative Solubility of a Solute in Two Solvents

1. (a) Which liquid is denser, decane or water? water

   (b) What experimental evidence supports your answer?
   The decane was on top and the water was on the bottom.

2. Color of iodine in water: clear

   Color of iodine in decane: pink

3. (a) In which of the two solvents used is iodine more soluble? decane

   (b) Cite experimental evidence for your answer.
   Water's color was unchanged, because the iodine did not dissolve into the water. In decane, the iodine mixed in and caused the color to change from clear to pink.

C. Miscibility of Liquids

1. Which liquid pairs tested are miscible?
   a. Kerosene and isopropyl alcohol c. water and isopropyl alcohol

2. How do you classify the liquid pair decane—H_2O, miscible or immiscible?
   immiscible

D. Rate of Dissolving Versus Particle Size

1. Time required for fine salt crystals to dissolve 17 seconds

2. Time required for coarse salt crystals to dissolve 63 seconds

3. Since the amount of salt, the volume of water, and the temperature of the systems were identical in both test tubes, how do you explain the difference in time for dissolving the fine vs. the coarse salt crystals?
   Fine crystals have more surface area than coarse crystals.

E. Rate of Dissolving Versus Temperature

1. Under which condition, hot or cold, did the salt dissolve faster? hot

2. Since the amount of salt, the volume of water, and the texture of the salt crystals were identical in both best tubes, how do you explain the difference in time for dissolving at the hot vs. cold temperatures?
   The molecules in hot water move faster than the molecules in cold water, which enables the water molecules to come into contact with the salt crystals, speeding up the rate of dissolution

F. Solubility vs. Temperature; Saturated and Unsaturated Solutions

Data Table: Circle the choices which best describe your observations.

	NaCl	NH_4Cl
1.0 g + 5 mL water	dissolved completely? yes/no saturated or unsaturated?	dissolved completely? yes/no saturated or unsaturated?
1.0 g + 5 mL water + 1.4 g	dissolved completely? yes/no saturated or unsaturated?	dissolved completely? yes/no saturated or unsaturated?
2.4 g + 5 mL water + heat	dissolved completely? yes/no saturated or unsaturated?	dissolved completely? yes/no saturated or unsaturated?
2.4 g + 5 mL water after cooling	dissolved completely? yes/no saturated or unsaturated?	dissolved completely? yes/no saturated or unsaturated?

G. Ionic Reactions in Solution

1. Write the word and formula equations representing the chemical reaction that occurred between the barium chloride solution, $BaCl_2(aq)$, and the sodium sulfate solution, $Na_2SO_4(aq)$.

Word Equation: Barium Chloride and Sodium Sulfate react to form Barium Sulfate and Sodium chloride.

Formula Equation:
$BaCl_2(aq) + Na_2SO_4(aq) \rightarrow BaSO_4(s) + NaCl(aq)$

2. (a) Which of the products is the white precipitate? $BaSO_4$

(b) What experimental evidence leads you to this conclusion?

There was a precipitate, therefore it is a precipitation reaction.

SUPPLEMENTARY QUESTIONS AND PROBLEMS

1. Use the solubility data in Table 9.1 to answer the following:
Show Calculations

(a) What is the percentage by mass of NaCl in a saturated solution of sodium chloride at 50°C?

$\frac{37.0 g}{100 g} \times 100$ 37.0%

(b) Calculate the solubility of potassium bromide at 23°C. Hint: Assume that the solubility increases by an equal amount for each degree between 20°C and 30°C.

65.2 / 20° = 70.6 / 30°

5.36 / 10 = 1.608 / 3

66.808 g KBr/100g water

(c) A saturated solution of barium chloride at 30°C contains 150 g water. How much additional barium chloride can be dissolved by heating this solution to 60°C?

38.2 / 100 = 57.3g / 150 at 30°

60° 46.6g / 100 = 69.9g / 150

69.9
− 57.3
12.6g more

12.6g BaCl2

2. A solution of KCl is saturated at 50°C.
Use Table 9.1

(a) How many grams of solute are dissolved in 100 g of water? 42.6g

(b) What is the total mass of the solution? 142.6g

(c) What is the mass percent of this solution at 50°C? 29.87%

42.6 / 142.6 (100)

(d) If the solution is heated to 100°C, how much more KCl can be dissolved in the solution without adding more water?

55.6 − 42.6 13g

(e) If the solution is saturated at 100°C and then cooled to 30°C, how many grams of solute will precipitate out?

55.6 − 37.0 18.6g

EXPERIMENT 10

Composition of Potassium Chlorate

MATERIALS AND EQUIPMENT

Solids: Reagent Grade potassium chlorate ($KClO_3$) and potassium chloride (KCl). **Solutions:** dilute (6 M) nitric acid (HNO_3) and 0.1 M silver nitrate ($AgNO_3$). Two No. 0 crucibles with covers; Ceramfab pad.

DISCUSSION

The **percentage composition** of a compound is the percentage by mass of each element in the compound. If the formula of a compound is known, the percentage composition can be calculated from the molar mass and the total mass of each element in the compound. The **molar mass** of a compound is determined by adding up the atomic masses of all the atoms making up the formula. The **total mass** of an element in a compound is determined by multiplying the atomic mass of that element by the number of atoms of that element in the formula. The percentage of each element is then calculated by dividing its total mass in the compound by the molar mass of the compound and multiplying by 100.

The percentage composition of many compounds may be directly determined or verified by experimental methods. In this experiment the percentage composition of potassium chlorate will be determined both experimentally and from the formula.

When potassium chlorate is heated to high temperatures (above 400°C) it decomposes to potassium chloride and elemental oxygen, according to the following equation:

$$2\,KClO_3(s) \xrightarrow{\Delta} 2\,KCl(s) + 3\,O_2(g)$$

The relative amounts of oxygen and potassium chloride are measured by heating a weighed sample of potassium chlorate until all of the oxygen has been released from the sample. This is accomplished when the sample is heated to constant weight. In this experiment you will heat, cool, and weigh the sample at least twice. If the sample loses more than 0.05 g after the second heating it has not been heated to constant weight and should be heated a third time.

From the experiment we obtain the following three values:

1. Mass of original sample ($KClO_3$).

2. Mass lost when sample was heated (Oxygen).

3. Mass of residue (KCl).

From these experimental values (and a table of atomic masses) we can calculate the following:

4. Percentage oxygen in sample (Experimental value)

$$= \left(\frac{\text{Mass lost by sample}}{\text{Original sample mass}}\right)(100)$$

5. Percentage KCl in sample (Experimental value)

$$= \left(\frac{\text{Mass of residue}}{\text{Original sample mass}}\right)(100)$$

6. Percentage oxygen in $KClO_3$ from formula (Theoretical value)

$$= \left(\frac{\text{3 at. masses of oxygen}}{\text{Molar mass of } KClO_3}\right)(100) = \left(\frac{3 \times 16.00\,\text{g}}{122.6\,\text{g}}\right)(100)$$

7. Percentage KCl in $KClO_3$ from formula (Theoretical value)

$$= \left(\frac{\text{Molar mass of KCl}}{\text{Molar mass of } KClO_3}\right)(100) = \left(\frac{74.55\,\text{g}}{122.6\,\text{g}}\right)(100)$$

8. Percentage error in experimental oxygen determination

$$= \left(\frac{\text{Theoretical value} - \text{Experimental value}}{\text{Theoretical value}}\right)(100)$$

PROCEDURE

PRECAUTIONS: Since potassium chlorate is a strong oxidizing agent it may cause fires or explosions if mixed or heated with combustible (oxidizable) materials such as paper. Observe the following safety precautions when working with potassium chlorate:

1 **Wear protective glasses.**

2. Use clean crucibles that have been heated and cooled prior to adding potassium chlorate.

3. Use Reagent Grade potassium chlorate.

4. **Dispose of any excess or spilled potassium chlorate as directed by your instructor. (Potassium chlorate may start fires if mixed with paper or other solid wastes.)**

5. Heat samples slowly and carefully to avoid spattering molten material—and to avoid poor experimental results.

NOTES:

1. Make all weighings to the highest precision possible with the balance available to you. Use the same balance to make all weighings for a given sample. Record all data directly on the report sheet as they are obtained.
2. Duplicate samples of potassium chlorate are to be analyzed, if two crucibles are available.
3. For utmost precision, handle crucibles with tongs after the initial heating.

A. Determining Percentage Composition

Place a clean, dry crucible (uncovered) on a clay triangle and heat for 2 or 3 minutes at the maximum flame temperature. The tip of the sharply defined inner-blue cone of the flame should almost touch and heat the crucible bottom to redness. Allow the crucible to cool. If two crucibles are being used, carefully transfer the first to a Ceramfab pad and heat the second while the first crucible is cooling.

Weigh the cooled crucible and its cover; add between 1 and 1.5 g of potassium chlorate; weigh again.

> **NOTE:** The crucible must be covered when potassium chlorate is being heated in it.

Place the covered crucible on the clay triangle and **heat gently for 8 minutes** with the tip of the inner-blue cone of the flame 6 to 8 cm (about 2.5 to 3 in.) below the crucible bottom. Then carefully lower the crucible or raise the burner until the tip of the sharply defined inner-blue cone just touches the bottom of the crucible, and heat for an additional 10 minutes. The bottom of the crucible should be heated to a dull red color during this period.

Grasp the crucible just below the cover with the concave part of the tongs and very carefully transfer it to a Ceramfab pad. Allow to cool (about 10 minutes) and weigh. Begin analysis of a second sample while the first is cooling.

After weighing, reheat the first sample for an additional 6 minutes at the maximum flame temperature (bottom of the crucible heated to a dull red color); cool and reweigh. If the residue is at constant weight, the last two weighings should be in agreement. If the mass decreased more than 0.05 g between these two weighings, repeat the heating and weighing until two successive weighings agree within 0.05 g. Use the final weight in your calculations.

Complete the analysis of the second sample following the same procedure used for the first.

B. Qualitative Examination of Residue

This part of the experiment should be started as soon as the final heating and weighing of the first sample is completed and while the second sample is in progress.

Number and place three clean test tubes in a rack. Put a pea-sized quantity of potassium chloride into tube No. 1 and a like amount of potassium chlorate into tube No. 2. Add 10 mL of distilled water to each of these two tubes and shake to dissolve the salts. Now add distilled water to the crucible containing the residue from the first sample so it is one-half full. Heat the uncovered crucible very gently for about 1 minute; transfer 1 to 2 mL of the resulting solution from the crucible to tube No. 3; add about 10 mL of distilled water and mix.

Test the solution in each tube as follows: Add 5 drops of dilute (6 M) nitric acid and 5 drops of 0.1 M silver nitrate solution. Mix thoroughly. Record your observations. This procedure using nitric acid and silver nitrate is a general test for chloride ions. The formation of a white precipitation is a positive test and indicates the presence of chloride ions. A positive test is obtained with any substance that produces chloride ions in solution.

Dispose of solutions and precipitates containing silver in the heavy metal waste container provided. Dispose of the remaining contents in the crucible down the sink.

NAME Ellen Jaggers

SECTION ________ DATE 10-23-14

REPORT FOR EXPERIMENT 10

INSTRUCTOR Prof Jaggi

Composition of Potassium Chlorate

A. Determining Percentage Composition

	Sample 1	Sample 2
1. Mass of crucible + cover	44.86	43.79g
2. Mass of crucible + cover + sample before heating	46.38	45.24g
3. Mass of crucible + cover + residue after 1st heating	46.38	44.69g
4. Mass of crucible + cover + residue after 2nd heating	46.38	44.69g
5. Mass of crucible + cover + residue after 3rd heating (if necessary)	KCl used	
6. Mass of original sample	by	1.45g
7. Mass lost (total) during heating	mistake	.55g
8. Final mass of residue		.90g
9. Experimental percent oxygen in sample ($KClO_3$)		38%
10. Experimental percent KCl in sample ($KClO_3$)		62%

6. Show sample 2 calculation setup: 45.24 − 43.79 = 1.45

7. Show sample 2 calculation setup: 45.24 − 44.69 = .55

8. Show sample 2 calculation setup: 1.45 − .55 = .9

9. Show sample 2 calculation setup: $\frac{.55}{1.45} \times 100 = 37.9$ round to 2 sig figs

10. Show sample 2 calculation setup: $\frac{.90}{1.45} \times 100 = 62\%$

11. Theoretical percent oxygen in $KClO_3$ 39%

Show calculation setup:

$$\frac{(16 \times 3)}{39 + 35 + (16 \times 3)} \times 100 = 39\%$$

12. Theoretical percent KCl in $KClO_3$ 61%

Show calculation setup

$$\frac{(39 + 35)}{(39 + 35) + (16 \times 3)} \times 100 = 61\%$$

13. Percent error in experimental % oxygen determination — Sample 2: 2.6%

Show sample 2 calculation setup

$$\frac{39 - 38}{39} \times 100 = 2.56$$ rounded to 2 sig figs

 NAME Ellen Jaggers

B. Qualitative Examination of Residue

1. Record what you observed when silver nitrate was added to the following:

   (a) Potassium chloride solution white precipitate

   (b) Potassium chlorate solution no change

   (c) Residue solution white precipitate

2. (a) What evidence did you observe that would lead you to believe that the residue was potassium chloride? The white precipitate formed, indicating the presence of chloride ions.

   (b) What would happen if you added silver nitrate to a solution of sodium chloride? Explain your answer. White precipitate would form, due to the presence of chloride in sodium chloride.

   (c) Did the evidence obtained in the silver nitrate tests of the three solutions prove conclusively that the residue actually was potassium chloride? Explain? No, that evidence alone only proves the presence of chloride ions, which are present in other substances as well.

QUESTIONS AND PROBLEMS

1. A student forgot to read the label on the jar carefully and put potassium chloride in the crucible instead of potassium chlorate. How would the results turn out? There would be no change in mass after heating. This actually did happen in sample 1.

2. What if a potassium chlorate sample is contaminated with KCl. Would the experimental % oxygen be higher or lower than the theoretical % oxygen? Explain your answer.
The experimental % oxygen would be lower, since KCl does not have any oxygen. The extra mass of KCl added to the $KClO_3$ would cause the ratio of oxygen to substance to be smaller, by increasing the denominator.

3. What if a potassium chlorate sample is contaminated with moisture. Would an analysis show the experimental % oxygen higher or lower than the theoretical % oxygen? Explain your answer? It would be lower, because the moisture would add to the mass of the $KClO_3$. Thus, in the calculation, the denominator would be greater, causing a lower percentage value.

4. Calculate the percentage of Cl in $Al(ClO_3)_3$ 10.6%

$$\frac{\text{molar mass of Cl}}{\text{molar mass of } Al(ClO_3)_3} \quad \frac{35}{3(27+35+(16\times3))} \quad \frac{35}{330}\times 100$$

5. Other metal chlorates when heated show behavior similar to that of potassium chlorate yielding metal chlorides and oxygen. Write the balanced formula equation for the reaction to be expected when calcium chlorate, $Ca(ClO_3)_2$ is heated.

$$2Ca(ClO_3)_2 \xrightarrow{\Delta} 2CaCl + 3O_2$$

EXPERIMENT 13

Ionization-Electrolytes and pH

MATERIALS AND EQUIPMENT

Demonstration. **Solids:** sodium chloride (NaCl) and sugar ($C_{12}H_{22}O_{11}$). **Liquid:** glacial acetic acid ($HC_2H_3O_2$). **Solutions:** 0.1 M ammonium chloride (NH_4Cl), 1 M ammonium hydroxide (NH_4OH), 1 M acetic acid ($HC_2H_3O_2$), saturated barium hydroxide [$Ba(OH)_2$], 0.1 M copper(II) sulfate ($CuSO_4$), 1 M hydrochloric acid (HCl), 0.1 M nickel(II) nitrate [$Ni(NO_3)_2$], 0.1 M sodium bromide (NaBr), 1 M sodium hydroxide (NaOH), 0.1 M sodium nitrate ($NaNO_3$), and dilute (3 M) sulfuric acid (H_2SO_4). Conductivity apparatus; magnetic stirrer and stirring bar.

Solids: calcium hydroxide [$Ca(OH)_2$], calcium oxide (CaO), iron wire (paper clips), magnesium ribbon (Mg), magnesium oxide (MgO), marble chips ($CaCO_3$), sodium bicarbonate ($NaHCO_3$), sulfur (S), and wood splints. **Solutions:** dilute (6 M) acetic acid ($HC_2H_3O_2$), dilute (6 M) ammonium hydroxide (NH_4OH), dilute (6 M) hydrochloric acid (HCl), dilute (6 M) nitric acid (HNO_3), phenolphthalein, 10 percent sodium hydroxide (NaOH), and dilute (3 M) sulfuric acid (H_2SO_4). 0.001 M HCl, 0.01 M HCl, 0.1 M HCl for pH measurements, pH meter.

DISCUSSION

A. Electrolytes

Pure water will not conduct an electric current. However, when many solutes are dissolved in water, the resulting aqueous solutions will conduct electricity. These solutes, called **electrolytes,** form ions which are free to move in the solution. The electrical current through the solution is the movement of these ions to the positive and negative electrodes. Electrolytes are **acids, bases, and salts,** depending on the ions in solution. Other substances such as sugar and alcohol dissolve in water but are nonconductors because they do not form ions and are called **nonelectrolytes**.

The ions in an aqueous electrolyte solution are the result of the **dissociation** or **ionization** of compounds in water. Compounds that dissociate or ionize in water are **acids, bases and salts.** For example:

Dissociation of NaOH (a base) and NaCl (a salt):

$$NaOH(s) \xrightarrow{H_2O} Na^+(aq) + OH^-(aq)$$

$$NaCl(s) \xrightarrow{H_2O} Na^+(aq) + Cl^-(aq)$$

Ionization of HCl (a strong acid) and $HC_2H_3O_2$ (a weak acid)

$$HCl(g) + H_2O(l) \longrightarrow H_3O^+(aq) + Cl^-(aq)$$

$$HC_2H_3O_2(l) + H_2O(l) \rightleftharpoons H_3O^+(aq) + C_2H_3O_2^-(aq)$$

The necessity for water in this ionization process is illustrated by the fact that, when hydrogen chloride is dissolved in benzene, no ions are formed and the solution is a nonconductor (nonelectrolyte).

Electrolytes are classifed as strong or weak depending on the extent to which they exist as ions in solutions. **Strong electrolytes** are essentially 100 percent ionized in water, that is they exist totally as ions in solution. **Weak electrolytes** are considerably less ionized, only a small amount of the dissolved substance exists as ions, the remainder being in the un-ionized or molecular from. Most salts are strong electrolytes; acids and bases occur as both strong and weak electrolytes. Examples are as follows:

Strong Electrolytes	**Weak Electrolytes**
Most salts	$HC_2H_3O_2$
HCl	H_2SO_3
H_2SO_4	HNO_2
HNO_3	H_2CO_3
$NaOH$	H_2S
KOH	$H_2C_2O_4$
$Ba(OH)_2$	H_3PO_4
$Ca(OH)_2$	NH_4OH

In the first part of this experiment, the conductivity of many aqueous solutions will be demonstrated.

B. Acids

1. **Acids** are described as substances that yield hydrogen ions (H^+) when dissolved in water. This definition was first proposed by the Swedish chemist Arrhenius (over 100 years ago) for electrolytes which share common properties such as sour taste and the ability to change the color of the plant dye, litmus to red. This definition is the simplest way to think of acids and still applies.

Many compounds can be recognized as acids from their written formulas. The ionizable hydrogen atoms, which are responsible for the acidity, are written first, followed by the symbols of the other elements in the formula. Examples are:

HCl	Hydrochloric acid	H_2CO_3	Carbonic acid
HNO_3	Nitric Acid	HNO_2	Nitrous acid
H_2SO_4	Sulfuric Acid	H_2SO_3	Sulfurous Acid
$HC_2H_3O_2$	Acetic Acid	$H_2C_2O_4$	Oxalic Acid
H_3PO_4	Phosphoric Acid		

Acids are formed by the reaction of nonmetallic oxides called **acid anhydrides** with water. For example:

$$SO_3(g) + H_2O(l) \longrightarrow H_2SO_4(aq)$$

The chemical properties of acids will be observed in Procedure B.

2. **Bronsted-Lowry Acids and Bases**

The more inclusive Bronsted-Lowry acid-base theory defines acids as proton (H^+) donors and bases as proton acceptors. Thus, water behaves as both an acid and a base, as illustrated by the equation:

$$\underset{\text{acid}}{H_2O} + \underset{\text{base}}{H_2O} \rightleftharpoons \underset{\text{acid}}{H_3O^+} + \underset{\text{base}}{OH^-}$$

One water molecule has donated a proton, H^+, (acted as an acid) and another water molecule has accepted a proton (acted as a base). The hydronium ion, H_3O^+, is a hydrated hydrogen ion (H^+H_2O). To simplify writing equations, the formula of the hydronium ion is often abbreviated H^+. However, free hydrogen ions do not actually exist in aqueous solutions.

C. Bases

The Arrhenius definition for **bases** describes them as substances that yield hydroxide ions ($OH-$) in water solutions. Bases change the color of litmus to blue. Common bases can be recognized by their formulas as a hydroxide ion (OH^-) combined with a metal or other positive ion. Examples are:

$NaOH$	Sodium hydroxide	KOH	Potassium hydroxide
$Ca(OH)_2$	Calcium hydroxide	$Mg(OH)_2$	Magnesium hydroxide
NH_4OH	Ammonium hydroxide		

The terms **alkali** and **alkaline** solutions are used synonymously with base and basic solutions.

Metal oxides that react with water to form bases are **basic anhydrides.** For example:

$$CaO(s) + H_2O(l) \longrightarrow Ca(OH)_2(aq)$$

The physical and chemical properties of bases will be observed in Procedure C.

D. Salts

Salts consist of a positively charged ion (H^+ excluded) and a negatively charged ion (O^{2-} and OH^- excluded). Salts may be formed by the reaction of acids and bases, or by replacing the hydrogen atoms in an acid with a metal, or by interaction of two other salts. There are many more salts than acids and bases. For example, for a single acid such as HCl we can produce many chloride salts (e.g. $NaCl$, KCl, $RbCl$, $CaCl_2$, NH_4Cl, $FeCl_3$, etc.)

The reaction of an acid and a base to form water and a salt is known as **neutralization.** For example:

$$HCl(aq) + NaOH(aq) \longrightarrow H_2O(l) + NaCl(aq)$$

E. The Importance and Measurement of H^+ Ion Concentration

An aqueous solution will be acidic, basic, or neutral, depending on the relative concentrations of H^+ and OH^-. In acidic solutions the concentration of the H^+ ions is greater than that of the OH^- ions. In basic solutions the concentration of the OH^- ions is greater than that of the H^+ ions. If the concentrations of H^+ and OH^- are equal (as in water), the solution is **neutral.**

There are two general methods for determining the relative concentrations of H^+ and OH^- and thus whether a solution is acid, alkaline, or neutral.

1. **Indicators** are organic compounds that change color at a particular hydrogen or hydroxide ion concentration. For example, litmus, a vegetable dye, shows a pink color in acidic solutions and a blue color in alkaline solutions. Another common indicator is phenolphthalein; it is colorless in acid solutions and pink in basic solutions. An indicator can only determine the relative concentrations of H^+ and OH^- within the range of its color changes.

2. A **pH meter** is an instrument designed so that it measures the H^+ directly and is used when an accurate measurement of the concentration of H^+ is needed. The pH meter is described and explained more fully in section (F).

F. Measuring pH

The H^+ ion has a great effect on many chemical reactions, including biological processes that sustain life. For example, the H^+ concentration of human blood is regulated to very close tolerances. The concentration of this important ion is expressed as pH rather than other concentration expressions such as molarity. The pH is defined by this formula:

$$\text{pH} = -\log[H^+]$$

The H^+ written in brackets $[H^+]$ represents the concentration of H^+ in moles/liter. The logarithm (log) of a number is simply the power to which 10 must be raised to give that number. Thus, the log of 0.001 is −3 ($0.001 = 10^{-3}$). Since pH is defined as the negative log of an $[H^+]$ value, then the pH of a solution with $[H^+] = 0.001$ moles/liter is $-(-3)$ or pH = 3.

The pH of pure water is 7.0 at 25°C and is said to be neutral, that is, it is neither acidic nor basic because $[H^+]$ and $[OH^-]$ are equal (10^{-7} moles/liter). Solutions that are acidic have pH values less than 7.0. Solution that are basic have pH values greater than 7.0.

$\text{pH} < 7.0$	acid solutions
$\text{pH} = 7.0$	neutral solutions
$\text{pH} > 7.0$	basic (alkaline) solutions

A pH meter is a delicate instrument that comes in many versions and sizes which share some of these common features.

1. pH meters use a glass electrode that is immersed in the solution being tested. The electrode converts the H^+ concentration into an electrical potential which is read by a voltmeter calibrated in pH units.

2. pH meters are calibrated against standard solutions of known pH. After calibration, the electrode is immersed in the test solutions and the pH meter provides a value relative to the standard.

3. The operation of pH meters varies with different models. Your instructor will demonstrate the use of the pH meter available to you.

PROCEDURE

Wear protective glasses.

A. Conductivity of Solutions—Instructor Demonstration

All of the following tests (except number 8) are performed in 18 × 150 mm test tubes, using the conductivity apparatus shown in Figure 13.1 or other suitable conductivity apparatus. The electrodes should be rinsed thoroughly with distilled water between the testing of different solutions.

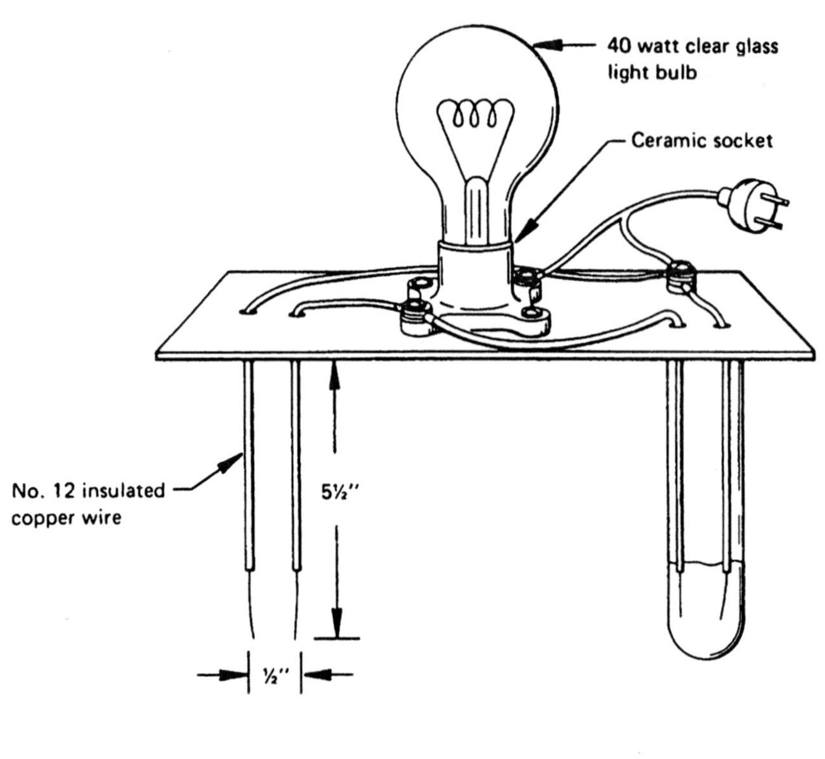

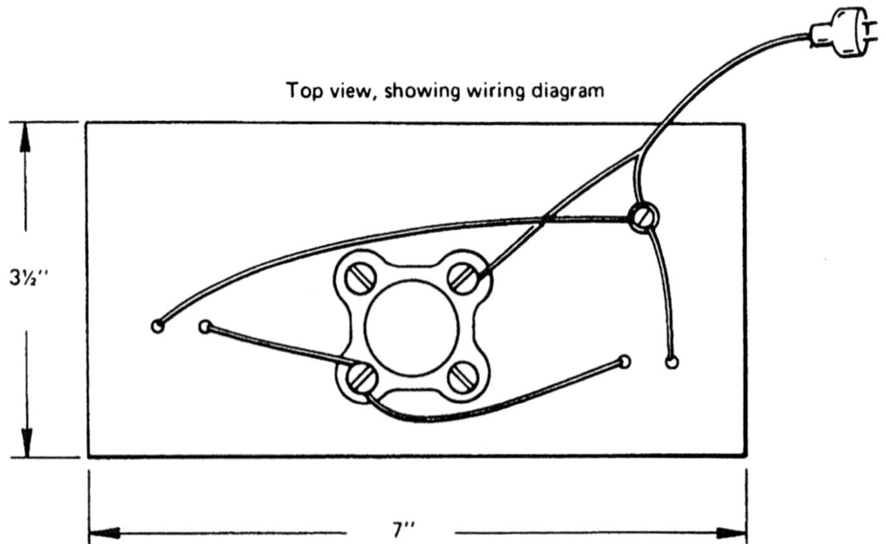

Figure 13.1 Conductivity apparatus

Each test is performed by filling a test tube about half full of the liquid to be tested, then raising the test tube up around a pair of electrodes. When a measurable number of ions are in solution, the solution will conduct the electric current and the light will glow. A dimly glowing light indicates a relatively small number of ions in solution; a brightly glowing light indicates a relatively large number of ions in solution.

NOTE: The student should complete the data table in the report form at the time the demonstration is performed.

1. Test the conductivity of distilled water.

2. Test the conductivity of tap water.

3. Add a small amount of sugar to a test tube that is half full of distilled water. Dissolve the sugar and test the solution for conductivity.

4. Add a small amount of sodium chloride to a test tube that is half full of distilled water. Dissolve the salt and test the solution for conductivity.

5. Remove the plug from the electrical outlet, clean and dry the electrodes, and reconnect the plug.

   (a) Test the conductivity of glacial acetic acid.

   (b) Pour out half of the acid, replace with distilled water, mix, and test the solution for conductivity.

   (c) Pour out half of the solution in 5(b), replace with distilled water, mix, and test the solution for conductivity.

6. Strong and weak acids and bases. Test the following 1 molar solutions for conductivity: (a) acetic acid, (b) hydrochloric acid, (c) ammonium hydroxide, (d) sodium hydroxide. If the conductivity apparatus has two sets of electrodes, as shown in Figure 13.1, the relative conductivity of the strong and weak acids or bases may be compared by alternately raising a tube of each solution around the electrodes. Clean and dry the electrodes (See No. 5).

7. Test the following 0.1 M salt solutions for conductivity: (a) sodium nitrate, (b) sodium bromide, (c) nickel(II) nitrate, (d) copper(II) sulfate, and (e) ammonium chloride.

8. Clean the electrodes well. Place about 25 mL of distilled water and 1 drop of dil. (3 M) sulfuric acid in a 150 mL beaker. Place the beaker on a magnetic stirrer and dip one set of electrodes into the solution. With the stirrer slowly turning, add saturated barium hydroxide solution dropwise until the light goes out completely. Add a few more milliliters of barium hydroxide solution.

B. Properties of Acids

Dispose of all solutions in the sink and flush with water. Take care to make sure that solids such as metal strips, splints, and unreacted marble chips do not go into the sink. They should be put into the wastebasket.

1. **Reaction with a Metal**

   (a) Into four consecutive test tubes place about 5 mL of dil. (6 M) hydrochloric, (3 M) sulfuric, (6 M) nitric, and (6 M) acetic acids.

   (b) Place a small strip of magnesium ribbon into each tube, one at a time, and test the gas evolved for hydrogen by bringing a burning splint to the mouth of the tube. If the liberation of gas is slow, stopper the test tube loosely for a minute or two before testing for hydrogen.

2. **Measurement of Acidity and pH**

   (a) Test dilute solutions of hydrochloric acid, acetic acid, and sulfuric acid by placing a drop of each acid from a stirring rod onto a strip of red and onto a strip of blue litmus paper. Note any color changes.

   (b) Add 2 drops of phenolphthalein solution to about 5 mL of distilled water. Add several drops of dilute hydrochloric acid, mix, and note any color change.

   (c) Use the pH meter to measure the pH of three dilutions of hydrochloric acid in this order: 0.001 M HCl, 0.01 M HCl, and 0.1 M HCl. *Rinse the electrodes thoroughly with distalled water when done.*

3. **Reaction with Carbonates and Bicarbonates**

   (a) Cover the bottom of a 150 mL beaker with a small quantity of sodium bicarbonate powder. Now add about 4 to 5 mL of dil. (6 M) hydrochloric acid to the beaker and cover with a glass plate. After about 30 seconds lower a burning splint into the beaker and observe the results. Dispose of the reaction mixture in the sink.

   (b) Repeat the above experiment, using a few granules of marble chips (calcium carbonate) instead of sodium bicarbonate. Allow the reaction to proceed for 2 minutes before testing with the burning splint. Dispose of unreacted marble chips in the wastebasket, not the sink.

4. **Reaction with Bases—Neutralization.** To about 25 mL of water in a beaker, add 3 drops of phenolphthalein solution and 5 drops of dil. (6 M) hydrochloric acid. Using a medicine dropper, add 10 percent sodium hydroxide solution dropwise, stirring after each drop, until the indicator in the solution changes color. Then add dilute hydrochloric acid, drop by drop, stirring after each drop, until the indicator becomes colorless again. Repeat the additions of base and acid one or two more times. Dispose of all solutions in the sink.

5. **Nonmetal Oxide plus Water** 

Dispose of all solutions in the sink.

   (a) **Do this part in the fume hood.** Place a small lump of sulfur in a deflagrating spoon and start it burning by heating in the burner flame. Lower the burning sulfur into a wide-mouth bottle containing 15 mL of distilled water and let the sulfur burn for 2 minutes. Remove the deflagrating spoon and quench the excess burning sulfur in a beaker of water. Cover the bottle with a glass plate and shake the bottle back and forth to dissolve the sulfur dioxide gas. Test the solution with blue litmus paper.

   (b) As shown in Figure 13.2, fit a test tube with a one-hole stopper containing a glass delivery tube long enough to extend to the bottom of another test tube. Place several pieces of marble chips and a few milliliters of dil. (6 M) hydrochloric acid into the tube and insert the stopper. Bubble the liberated carbon dioxide into another test tube containing 10 mL water, 1 drop of 10 percent sodium hydroxide solution, and 2 drops of phenolphthalein solution. Record the results.

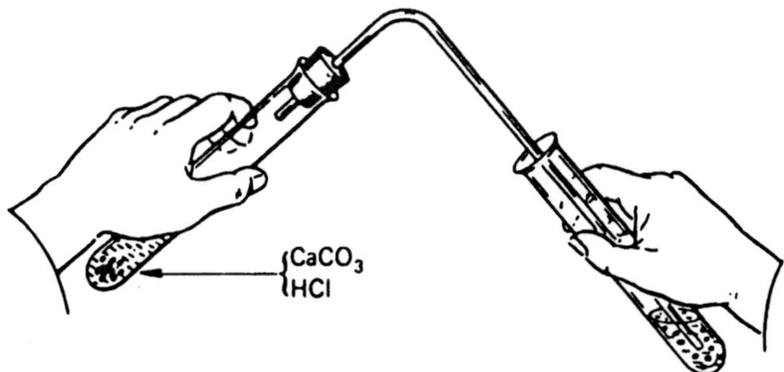

Figure 13.2 Generator for carbon dioxide

C. Properties of Bases

Dispose of all solutions in the sink.

1. "Feel" Test. Make very dilute solutions by adding 5 drops of dilute (6 M) ammonium hydroxide to 10 mL of water in a test tube and 3 drops of 10 percent sodium hydroxide solution to 10 mL of water in another test tube. Rub a small amount of each very dilute solution between your fingers to obtain the characteristic "feel" of a hydroxide (base) solution. Wash your hands thoroughly immediately after making the "feel" test. Save the very dilute base solutions for the measurement of pH in the next section, C2(c).

2. **Measurement of Alkalinity**

(a) Test the two base (alkaline) solutions prepared in C.1 with both red and blue litmus paper. Note any color changes.

(b) Add 2 drops of phenolphthalein solution to each of the two alkaline solutions prepared in C.1. Note any color changes.

(c) Pour the dilute ammonium hydroxide and sodium hydroxide that were prepared in the previous step into separate small beakers. Use the pH meter to measure the pH of these alkaline solutions. *Dip the electrode in dilute acetic acid and rinse thoroughly with distilled water when done.*

3. **Metal Oxides plus Water**

(a) Place 10 mL of water and 2 drops of phenolphthalein solution in each of 3 test tubes. Add a pinch of calcium oxide to the first, magnesium oxide to the second, and calcium hydroxide to the third tube. Note and record the results.

(b) Wind the end of a 5 cm piece of iron wire (or paper clip) around a small marble chip. Grasp the wire with tongs and heat the marble chip in the hottest part of the burner (flame for about 2 minutes—the edges of the chip should become white hot while being heated. Allow the chip to cool; then drop it into a beaker containing 15 mL of water and 2 drops of phenolphthalein solution. For comparison, repeat this part of the experiment with a marble chip which has not been heated. Note the results. **Return the iron wire to the reagent shelf.** Dispose marble chips in the wastebasket.

4. **Reaction with Acids—Neutralization.** Review Part B.4.

NAME ______________________

SECTION __________ DATE __________

REPORT FOR EXPERIMENT 13

INSTRUCTOR ______________________

Ionization—Electrolytes and pH

A. Conductivity of Solutions—Instructor Demonstration

Complete the table for each of the substances tested in the ionization demonstration. Place an "X" in the column where the property of the substance tested fits the column description.

	Nonelectrolyte	Strong Electrolyte	Weak Electrolyte
1. Distilled Water			
2. Tap water			
3. Sugar			
4. NaCl			
5. a. $HC_2H_3O_2$ (glacial)			
b. 1st dilution			
c. 2nd dilution			
6. a. 1 M $HC_2H_3O_2$			
b. 1 M HCl			
c. 1 M NH_4OH			
d. 1 M NaOH			
7. a. $NaNO_3$			
b. NaBr			
c. $Ni(NO_3)_2$			
d. $CuSO_4$			
e. NH_4Cl			

8. (a) Write an equation for the chemical reaction that occurred between sulfuric acid and barium hydroxide.

   (b) Explain in terms of the properties of the products formed why the light went out when barium hydroxide was added to sulfuric acid solution, even though both of these reactants are electrolytes.

   (c) Explain why the light came on again when additional barium hydroxide was added.

9. In the conductivity tests, what controlled the brightness of the light?

10. Write an equation to show how acetic acid reacts with water to produce ions in solution.

11. What classes of compounds tested are electrolytes?

B. Properties of Acids

1. Reaction with a Metal

(a) Write the formulas of the acids which liberated hydrogen gas when reacting with magnesium metal.

____

(b) Write equations to represent the reactions in which hydrogen gas was formed.

2. Measurement of Acidity and pH

(a) What is the effect of acids on the color of red litmus?

(b) What is the effect of acids on the color of blue litmus?

(c) What color is phenolphthalein in an acid solution? ________

(d) What was the pH of the hydrochloric acids tested?

0.001 M ________ 0.01 M ________ 0.1 M ________

(e) Which pH measured has the highest number of H^+ in solution? ________

(f) What is the H^+ concentration in an acid with of pH 4.6?
Express your answer as a power of 10 ________

Refer to Study Aid 4 if you need help with using your calculator to convert the pH into H^+ concentration using the antilog function.

3. Reaction with Carbonates and Bicarbonates

(a) What gas is formed in these reactions?

Name ____________ Formula ____________

(b) What happened to the burning splint when it was thrust into the beaker?

(c) What do you conclude about one of the properties of the gas in the beaker, based on the behavior of the burning splint?

(d) Complete and balance the equations representing the reactions:

$NaHCO_3(s) + HCl(aq) \longrightarrow$

$CaCO_3(s) + \quad HCl(aq) \longrightarrow$

4. **Reaction of Acids with Bases—Neutralization**

(a) Write an equation for the neutralization reaction of HCl and NaOH.

(b) How did you know when all the acid was neutralized?

5. **Nonmetal Oxide plus Water**

(a) Write an equation for the combustion of sulfur in air.

(b) What acid is formed when the product of the sulfur combustion reacts with water?

Name ______________________ Formula ______________________

Write the equation for its formation.

(c) What evidence in this experiment leads you to believe that carbon dioxide in water has acidic properties?

(d) What acid is formed when carbon dioxide reacts with water?

Name ______________________ Formula ______________________

C. Properties of Bases

1. **"Feel" Test.** What is the characteristic feel of basic solutions?

2. **Measurement of Alkalinity**

(a) What is the effect of bases on the color of red litmus?

(b) What is the effect of bases on the color of blue litmus?

(c) What color is phenolphthalein in a basic solution? __________

(d) What was the pH for each dilute base tested?

$NH_4OH(aq)$ ____________ $NaOH(aq)$ ____________

(e) Which base tested has the highest number of H^+ in solution? __________

(f) What is the H^+ concentration in the strongest base tested? Express your answer as a power of 10. __________

Refer to Study Aid 4 if you need help using your calculator to convert the pH into H^+ concentration using the antilog function.

3. **Metal Oxides plus Water**

(a.1) Color (if any) produced by phenolphthalein.

Color with CaO in water __________

Color with MgO in water __________

Color with $Ca(OH)_2$ in water __________

(a.2) Complete and balance these equations:

$CaO(s) + \quad H_2O(l) \longrightarrow$

$MgO(s) + \quad H_2O(l) \longrightarrow$

(b.1) The formula for marble is $CaCO_3$, What compounds are formed when it is heated strongly?

(b.2) Write the equation representing this decomposition:

$CaCO_3(s) \xrightarrow{\Delta}$

(b.3) What evidence led you to formulate the composition of the solid residue after heating the marble chip?

ADDITIONAL QUESTIONS AND PROBLEMS

1. State whether each of the formulas below represents an **acid,** a **base,** a **salt,** an **acid anhydride,** a **basic anhydride,** or **none** of these types of compounds:

CuF_2	______________	$CaSO_4$	______________
$Ba(OH)_2$	______________	C_2H_4	______________
$LiOH$	______________	$C_{12}H_{22}O_{11}$	______________
$HBrO_3$	______________	HI	______________
$RaCO_3$	______________	P_2O_5	______________
KNO_2	______________	HCN	______________
$H_2C_2O_4$	______________	MgO	______________

2. Complete and balance the following equations and name the product formed. (Only one product is formed in each case.)

	Name of Product
(a) $K_2O(s) + H_2O(l) \longrightarrow$	______________
(b) $SrO(s) + H_2O(l) \longrightarrow$	______________
(c) $SO_3(s) + H_2O(l) \longrightarrow$	______________
(d) $N_2O_5(s) + H_2O(l) \longrightarrow$	______________

EXPERIMENT 14

Identification of Selected Anions

MATERIALS AND EQUIPMENT

Liquids: Decane ($C_{10}H_{22}$). **Solutions:** 0.1 M barium chloride ($BaCl_2$), freshly prepared chlorine water (Cl_2), dilute (6 M) hydrochloric acid (HCl), dilute (6 M) nitric acid (HNO_3), 0.1 M silver nitrate ($AgNO_3$), 0.1 M sodium bromide (NaBr), 0.1 M sodium carbonate (Na_2CO_3), 0.1 M sodium chloride (NaCl), 0.1 M sodium iodide (NaI), 0.1 M sodium phosphate (Na_3PO_4), 0.1 M sodium sulfate (Na_2SO_4), and unknown solutions. Wash bottle for distilled water.

DISCUSSION

The examination of a sample of inorganic material to identify the ions that are present is called **qualitative analysis.** To introduce qualitative analysis, we will analyze for six anions (negatively charged ions). The ions selected for identification are chloride (Cl^-), bromide (Br^-), iodide (I^-), sulfate ($SO_4{}^{2-}$), phosphate ($PO_4{}^{3-}$) and carbonate ($CO_3{}^{2-}$).

Qualitative analysis is based on the fact that no two ions behave identically in all of their chemical reactions. Identification depends on appropriate chemical tests coupled with careful observation of such characteristics as solution color, formation and color of precipitates, evolution of gases, etc. Test reactions are selected to identify the ions in the fewest steps possible. In this experiment only one anion is assumed to be present in each sample. If two or more anions must be detected in a single solution, the scheme of analysis can be considerably more complex.

Silver Nitrate Test

When solutions of the sodium salts of the six anions are reacted with silver nitrate solution, the following precipitates are formed: AgCl, AgBr, AgI, Ag_3PO_4, and Ag_2CO_3. Ag_2SO_4 is moderately soluble and does not precipitate at the concentrations used in these solutions. When dilute nitric acid is added, the precipitates Ag_3PO_4, and Ag_2CO_3 dissolve; AgCl, AgBr, and AgI remain undissolved. Acids react with carbonates to form CO_2 (g). Look for gas bubbles when nitric acid is added to the silver precipitates.

In some cases a tentative identification of an anion may be made from the silver nitrate test. This identification is based on the color of the precipitate and on whether or not the precipitate is soluble in nitric acid. However, since two or more anions may give similar results, second or third confirmatory tests are necessary for positive identification.

Barium Chloride Test

When barium chloride solution is added to solutions of the sodium salts of the six anions, precipitates of $BaSO_4$, $Ba_3(PO_4)_2$ and $BaCO_3$, are obtained. No precipitate is obtained with Cl^-, Br^-, or I^-.

When dilute hydrochloric acid is added, the precipitates $Ba_3(PO_4)_2$ and $BaCO_3$ dissolve; $BaSO_4$ does not dissolve. Look for CO_2 gas bubbles.

Organic Solvent Test

The silver nitrate test can prove the presence of a halide ion (Cl^-, Br^-, or I^-) because the silver precipitates of the other three anions dissolve in nitric acid. But the colors of the three silver halides do not differ sufficiently to establish which halide ion is present.

Adding chlorine water (Cl_2 dissolved in water) to halide salts in solution will oxidize bromide ion to free bromine (Br_2) and iodine ion to free iodine (I_2). The free halogen may be extracted from the water solution by adding an immiscible organic solvent such as decane and shaking vigorously. The colors of the three halogens in organic solvents are quite different. Cl_2 is pale yellow, Br_2 is yellow-orange to reddish-brown, and I_2 is pink to violet. After adding chlorine water and shaking, a yellow-orange to reddish-brown color in the decane layer indicates that Br^- was present in the original solution; a pink to violet color in the decane layer indicates that I^- was present. However, a pale yellow color does not indicate Cl^-, since Cl_2 was added as a reagent. But if the silver nitrate test gives a white precipitate that is insoluble in nitric acid, and the organic solvent test shows no Br^- or I^-, then you can conclude that Cl^- was present.

Though we have described many of the expected results of these tests, it is necessary to test known solutions to actually see the results of the tests and to develop satisfactory experimental techniques. During this experiment, you will perform these tests on six known anions.

Then, two "unknown" solutions, each containing one of the six anions, will be analyzed. When an unknown is analyzed, the results should agree in all respects with one of the known anions. If the results do not fully agree with one of the six known ions, either the testing has been poorly done or the unknown does not contain any of the specified ions.

Three different kinds of equations may be used to express the behavior of ions in solution. For example, the reaction of the chloride ion (from sodium chloride) may be written.

1. $NaCl(aq) + AgNO_3(aq) \longrightarrow AgCl(s) + NaNO_3(aq)$

2. $Na^+(aq) + Cl^-(aq) + Ag^+(aq) + NO_3^-(aq) \longrightarrow AgCl(s) + Na^+(aq) + NO_3^-(aq)$

3. $Cl^-(aq) + Ag^+(aq) \longrightarrow AgCl(s)$

Equation (1) is the **formula (un-ionized) equation;** it shows the formulas of the substances in the equation as they are normally written. Equation (2) is the **total ionic equation;** it shows the substances as they occur in solution. Strong electrolytes are written as ions; weak electrolytes, precipitates, and gases are written in their un-ionized or molecular form. Equation (3) is the **net ionic equation;** it includes only those substances or ions in Equation (2) that have undergone a chemical change. Thus Na^+ and NO_3^- (sometimes called the "spectator" ions) have not changed and do not appear in the net ionic equation. In both the total ionic and net ionic equations, the atoms and charges must be balanced.

PROCEDURE

Wear protective glasses

1. Clean eight test tubes and rinse each twice with 5 mL of distilled water. The first six test tubes are for the known solutions that will be tested to demonstrate the expected reactions with each anion. Use a marker to label these tubes as follows: NaCl, NaBr, NaI, Na_2SO_4, Na_3PO_4 and Na_2CO_3. The last two tubes are for your unknowns and should be left blank for now. Arrange these test tubes in order in your test tube rack.

2. Clean and rinse two more test tubes and take them to your instructor for your unknown solutions and their identification code. Label them with the code numbers immediately. To avoid possible confusion with the empty unknown test tubes in the rack, put these coded tubes aside in a beaker. Record the code of these unknowns in the top right-hand columns of your report form and label each of the blank tubes in the rack with one of these unknown code numbers.

Pour 2 mL (no more) of each of the six known solutions—one solution per tube—and 2 mL of the corresponding unknown into each unknown tube. Save the remaining portions of the unknown solutions for tests B and C.

You can save considerable time by measuring out 2 mL into the first test tube and using the height of this liquid in the test tube as a guide for measuring out the others.

Dispose of solutions containing decane in the container marked "Waste organic solvents." Dispose of solutions containing silver, and barium, in the "heavy metals waste" container.

> For each of the following tests that will be performed on known and unknown solutions, there is a corresponding block on the report form where observations should be recorded. If a precipitate forms, record "ppt formed" and include its color. If no precipitate forms, record "no ppt." When dissolving precipitates, record "ppt dissolved" or "ppt did not dissolve." For the decane solubility test, indicate the color of the decane layer.

A. Silver Nitrate Test

Silver nitrate will stain your skin black. If any silver nitrate gets on your hands, wash it off immediately to avoid these stains.

Add about 1 mL of 0.1 M silver nitrate solution to each test tube. Record the results. Now add about 3 mL of dilute (6 M) nitric acid to each test tube; stopper and shake well. Record the results.

B. Barium Chloride Test

Wash all eight test tubes and rinse each tube twice with distilled water. Again put about 2 mL of the specified solution into each of the eight test tubes. Add about 2 mL of 0.1 M barium chloride solution to each test tube and mix. Record the results. Now add 3 mL of dilute hydrochloric acid to each tube; stopper and shake well. Record the results.

C. Organic Solvent Test

Again wash and rinse all eight test tubes. Again put about 2 mL of the specified solution into each of the eight test tubes. Now add about 2 mL of decane and about 2 mL of chlorine water to each test tube; stopper and shake well. Record the results.

After completing the three tests, compare the results of the known solutions with your observations for your unknown solutions. Record the formula of the anion present in each solution on the report form (Part D).

NAME ____________________

SECTION __________ DATE __________

INSTRUCTOR ____________________

REPORT FOR EXPERIMENT 14

Identification of Selected Anions

	NaCl	NaBr	NaI	Na_2SO_4	Na_3PO_4	Na_2CO_3	Unknown No. ______	Unknown No. ______
A. $AgNO_3$ Test Addition of $AgNO_3$ solution								
Addition of dil. HNO_3								
B. $BaCl_2$ Test Addition of $BaCl_2$ solution								
Addition of dil. HCl								
C. Organic Solvent Test Color of decane layer								
D. Formula of anion present in the solution tested.								

QUESTIONS AND PROBLEMS

1. The following three solutions were analyzed according to the scheme used in this experiment. Which one, if any, of the ions tested, is present in each solution? If the data indicate that none of the six is present, write the word "None" as your answer.

   (a) **Silver Nitrate Test.** Yellow precipitate formed, which dissolved in dilute nitric acid.

   Barium Chloride Test. White precipitate formed, which dissolved in dilute hydrochloric acid.

   Organic Solvent Test. The decane layer remained almost colorless after treatment with chlorine water.

   Anion present ___________________

   (b) **Silver Nitrate Test.** Red precipitate formed, which dissolved in dilute nitric acid to give an orange solution.

   Barium Chloride Test. Yellow precipitate formed, which dissolved in dilute hydrochloric acid to give an orange solution.

   Organic Solvent Test. The decane layer remained almost colorless after treatment with chlorine water.

   Anion present ___________________

   (c) **Silver Nitrate Test.** Yellow precipitate formed, which did not dissolve in dilute nitric acid.

   Barium Chloride Test. No precipitate formed.

   Organic Solvent Test. The decane layer turned reddish-brown.

   Anion present ___________________

2. Write formula, total ionic, and net ionic equations for the following reactions: Use the solubility table in Appendix 5 for reactions that were not observed directly in this experiment. All reactions are in aqueous solutions.

   (a) Sodium bromide and silver nitrate.

   (b) Sodium carbonate and silver nitrate.

   (c) Sodium arsenate and barium chloride.

3. Write net ionic equations for the following reactions. Assume that a precipitate is formed in each case.

   (a) Sodium iodide and silver nitrate.

   (b) Sodium acetate and silver nitrate.

   (c) Sodium phosphate and barium chloride.

   (d) Sodium sulfate and barium chloride.

EXPERIMENT 17

Lewis Structures and Molecular Models

MATERIALS AND EQUIPMENT

Special equipment: Ball-and-stick molecular model sets

DISCUSSION

Molecules are stable groups of covalently bonded atoms, usually nonmetallic atoms. Chemists study models of molecules to learn more about their bonds, the spatial relationships between atoms and the shapes of molecules. Using models helps us to predict molecular structure.

A. Valence Electrons

Every atom has a nucleus surrounded by electrons which are held within a region of space by the attractive force of the positive protons in the nucleus. The electrons in the outermost energy level of an atom are called valence electrons. The **valence electrons** are involved in bonding atoms together to form compounds. For the representative elements, the number of valence electrons in the outermost energy level is the same as their group number in the periodic table (Groups 1A–7A). For example, sulfur in Group 6A has six valence electrons and potassium in Group 1A has one valence electron.

B. Lewis Structures

Lewis electron dot structures are a useful device for keeping track of valence electrons for the representative elements. In this notation, the nucleus and core electrons are represented by the atomic symbol and the valence electrons are represented by dots around the symbol. Although there are exceptions, Lewis structures emphasize an octet of electrons arranged in the noble gas configuration, ns^2np^6. Lewis structures can be drawn for individual atoms, monatomic ions, molecules, and polyatomic ions.

1. **Atoms and Monatomic Ions:** A Lewis structure for an atom shows its symbol surrounded by dots to represent its valence electrons. Monatomic ions form when an atom loses or gains electrons to achieve a noble gas electron configuration. The Lewis structure for a monatomic ion is enclosed by brackets with the charge of the ion shown. The symbol is surrounded by the valence electrons with the number adjusted for the electrons lost or gained when the ion is formed. This is the basis of ionic bond formation which is not included in this experiment.

Examples:	sulfur atom	sulfide ion	potassium atom	potassium ion
	$:\ddot{\underset{\cdot}{S}}\cdot$	$[:\ddot{\underset{\cdot\cdot}{S}}:]^{2-}$	$K\cdot$	$[K]^+$

2. **Molecules and Polyatomic Ions:** Lewis structures for molecules and polyatomic ions emphasize the principle that atoms in covalently bonded groups achieve the noble gas configuration, ns^2np^6. Since all noble gases except helium have eight valence electrons, this is

often called the octet rule. Although many molecules and ions have structures which support the octet rule, it is only a guideline. There are many exceptions. One major exception is the hydrogen atom which can covalently bond with only one atom and share a total of two electrons to form a noble gas configuration like helium. All of the examples in this experiment follow the octet rule except hydrogen.

A Lewis structure for covalently bonded atoms is a two-dimensional model in which one pair of shared electrons between two atoms is a single covalent bond represented by a short line; unshared or lone pairs of electrons are shown as dots. Sometimes two pairs of electrons are shared between two atoms forming a double bond and are represented by two short lines. It is even possible for two atoms to share three pairs of electrons forming a triple bond, represented by three short lines. For a polyatomic ion, the rules are the same except that the group of atoms is enclosed in brackets and the overall charge of the ion is shown. For example:

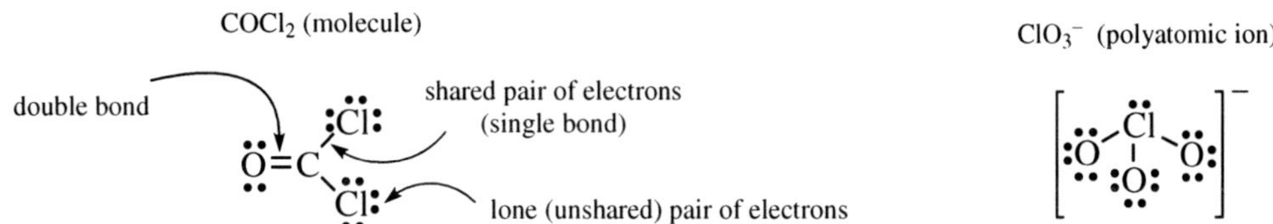

The rules for writing Lewis structures for molecules and polyatomic ions will be provided in the procedure section so you can use your Lewis structures to build three-dimensional models.

C. Molecular Model Building

The three-dimensional structure of a molecule is difficult to visualize from a two-dimensional Lewis structure. Therefore, in this experiment, a ball-and-stick model kit (molecular "tinker toys") is used to build models so the common geometric patterns into which atoms are arranged can be seen. Each model that is constructed must be checked by the instructor and described by its geometry and its bond angles on the report form.

D. Molecular Geometry

Atoms in a molecule or polyatomic ion are arranged into geometric patterns that allow their electron pairs to get as far away from each other as possible (which minimizes the repulsive forces between them). The theory underlying this molecular model is known as the valence shell electron pair repulsion **(VSEPR) theory.** All of the geometric structures in this experiment fall into the following patterns:

1. **Tetrahedral:** four pairs of shared electrons (no pairs of lone (unshared) electrons) around a central atom.

2. **Trigonal pyramidal:** three pairs of shared electrons and one pair of unshared electrons around a central atom.

3. **Trigonal planar:** three groups of shared electrons around a central atom. Two of these groups are single bonds and one group is a double bond made up of two pairs of shared electrons. There are no unshared electrons around the central atom.

4. **Bent:** two groups of shared electrons (in single or double bonds) and one or two pairs of unshared electrons around a central atom.

5. **Linear:** two groups of shared electrons, usually double bonds with two shared electron pairs between two atoms, and no unshared electrons around a central atom. When there are only two atoms in a molecule or ion, and there is no central atom (HBr, for example), the geometry is also linear. These patterns are described more extensively in Section E, which follows.

> **NOTE:** There are other electron arrangements and molecular geometries. Since they do not follow the octet rule, they are not included in this experiment.

E. Bond Angles

Bond angles always refer to the angle formed between two end atoms with respect to the central atom. If there is no central atom, there is no bond angle.

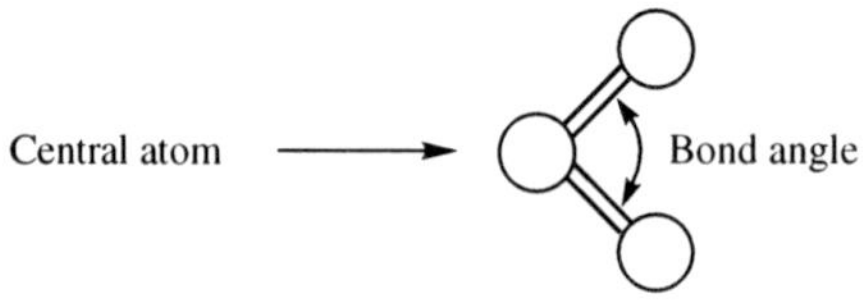

The size of the angle depends mainly on the repulsive forces of the electrons around the central atom. The molecular model kits are designed so that these angles can be determined when sticks representing electron pairs are inserted into pre-drilled holes.

1. Bond angles for atoms bonded to a central atom without unshared electrons on the central atom.

   a. For four pairs of shared electrons around a central atom (tetrahedral geometry) the angle between the bonds is approximately **109.5°**.

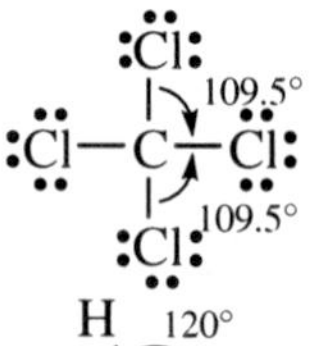

   b. For three atoms bonded to a central atom, (trigonal planar) the angle is **120°**. The shared electron pairs can be arranged in single or double bonds.

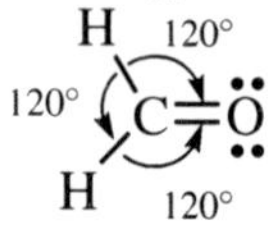

   c. For two atoms bonded to a central atom (linear) the angle is **180°**. The shared electrons are usually arranged in double bonds.

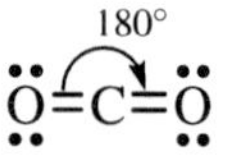

   d. Linear diatomic molecules or ions with no central atom do not have a bond angle.

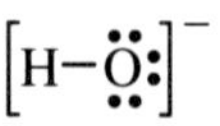

2. Bond angles for atoms bonded to a central atom **with** unshared electrons on the central atom.

When some of the valence electrons around a central atom are unshared, the VSEPR theory can be used to predict changes in spatial arrangements. An unshared pair of electrons on the central atom has a strong influence on the shape of the molecule. It reduces the angle of bonding pairs by squeezing them toward each other.

For example:

tetrahedral	**trigonal pyramidal**	**bent**	**bent**
No unshared electrons	1 unshared electron pair	2 unshared lone pairs	1 unshared pair
4 pairs shared electrons	3 pairs shared electrons	2 pairs shared electrons	2 groups shared electrons
CH_4	NH_3	H_2O	
109.5° 109.5° 109.5°	107° 107°	105°	119°

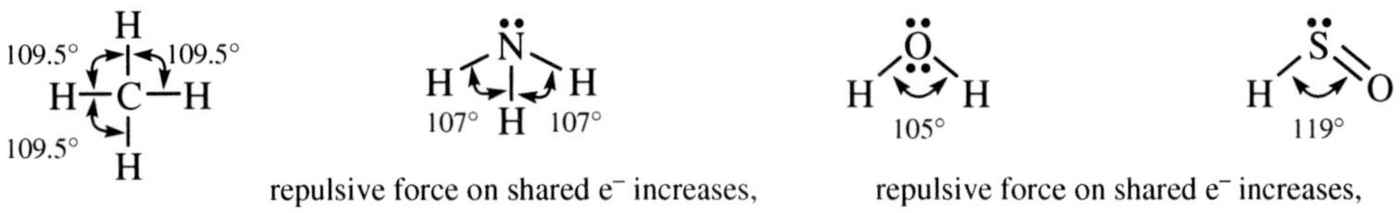

repulsive force on shared e^- increases, which pushes down on H atoms

repulsive force on shared e^- increases, which pushes down on peripheral atoms

F. Bond Polarity

Electrons shared by two atoms are influenced by the positive attractive forces of both atomic nuclei. For like atoms, these forces are equal. For example, in diatomic molecules such as H_2 or Cl_2 the bonded atoms have exactly the same electronegativity (affinity for the bonding electrons). Electronegativity values for most of the elements have been assigned.

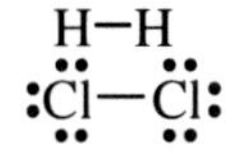

H—H

:C̤̈l—C̤̈l:

Electronegativity Table

9 — Atomic number
F — Symbol
4.0 — Electronegativity

1 H 2.1																	2 He
3 Li 1.0	4 Be 1.5											5 B 2.0	6 C 2.5	7 N 3.0	8 O 3.5	9 F 4.0	10 Ne
11 Na 0.9	12 Mg 1.2											13 Al 1.5	14 Si 1.8	15 P 2.1	16 S 2.5	17 Cl 3.0	18 Ar
19 K 0.8	20 Ca 1.0	21 Sc 1.3	22 Ti 1.4	23 V 1.6	24 Cr 1.6	25 Mn 1.5	26 Fe 1.8	27 Co 1.8	28 Ni 1.8	29 Cu 1.9	30 Zn 1.6	31 Ga 1.6	32 Ge 1.8	33 As 2.0	34 Se 2.4	35 Br 2.8	36 Kr
37 Rb 0.8	38 Sr 1.0	39 Y 1.2	40 Zr 1.4	41 Nb 1.6	42 Mo 1.8	43 Tc 1.9	44 Ru 2.2	45 Rh 2.2	46 Pd 2.2	47 Ag 1.9	48 Cd 1.7	49 In 1.7	50 Sn 1.8	51 Sb 1.9	52 Te 2.1	53 I 2.5	54 Xe
55 Cs 0.7	56 Ba 0.9	57–71 La-Lu 1.1–1.2	72 Hf 1.3	73 Ta 1.5	74 W 1.7	75 Re 1.9	76 Os 2.2	77 Ir 2.2	78 Pt 2.2	79 Au 2.4	80 Hg 1.9	81 Tl 1.8	82 Pb 1.8	83 Bi 1.9	84 Po 2.0	85 At 2.2	86 Rn
87 Fr 0.7	88 Ra 0.9	89–103 Ac–Lr 1.1–1.7	104 Rf —	105 Db —	106 Sg —	107 Bh —	108 Hs —	109 Mt —	110 Ds —	111 Rg —							

* The electronegativity value is given below the symbol of each element.

In general, electronegativity increases as we move across a period and up a group on the periodic table. Identical atoms with identical attractions for their shared electron pairs form **nonpolar covalent bonds.** Unlike atoms exert unequal attractions for their shared electrons and form **polar covalent bonds.**

Electronegativity is used to determine the direction of bond polarity which can be indicated in the Lewis structure by replacing the short line for the bond with a modified arrow (⟼) pointed towards the more electronegative atom. For example, nitrogen and hydrogen have electronegativity values of 3.0 and 2.1, respectively. The N—H bond is thus represented as

N⟷H with the arrow directed toward the more electronegative nitrogen atom. Then, the Lewis structure can be redrawn with arrows replacing the dashes as shown for NH_4^+ and NH_3.

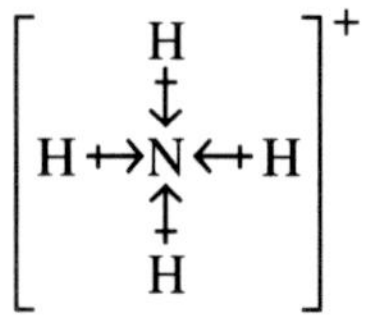

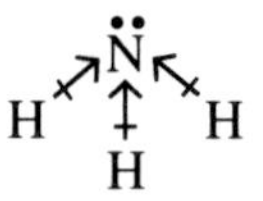

$$\left[\begin{array}{c} H \\ \downarrow \\ H \rightarrow N \leftarrow H \\ \uparrow \\ H \end{array}\right]^+ \qquad \begin{array}{c} \ddot{N} \\ H \nearrow \ \uparrow \ \nwarrow H \\ H \end{array}$$

G. Molecular Dipoles

When there are several polar covalent bonds within a molecule or a polyatomic ion such as in NH_3 and NH_4^+ the polar effect of these bonds around a central atom can be cancelled if they are arranged **symmetrically** as shown in CCl_4 below. On the other hand, if the arrangement of the polar bonds is asymmetrical, as in the bent water molecule, H_2O, the resulting molecule has a definite positive end and oppositely charged negative end, and the molecule is called a dipole. In water, the H atoms have a partial positive charge, δ^+ , and the O atom has a partial negative charge, δ^-. The symmetry, or lack of symmetry of molecules and polyatomic ions, can generally be seen in the three-dimensional model.

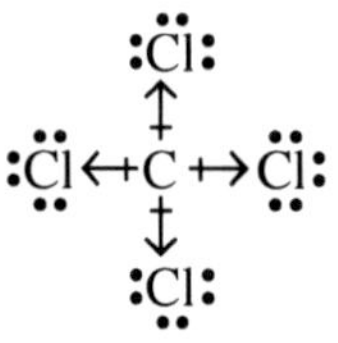

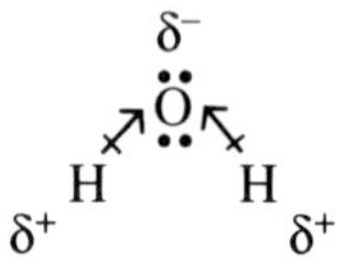

$$\begin{array}{c} :\ddot{Cl}: \\ \uparrow \\ :\ddot{Cl} \leftarrow C \rightarrow \ddot{Cl}: \\ \downarrow \\ :\ddot{Cl}: \end{array} \qquad \begin{array}{c} \delta^- \\ \ddot{O} \\ \nearrow \quad \nwarrow \\ H \qquad H \\ \delta^+ \qquad\quad \delta^+ \end{array}$$

PROCEDURE

Follow steps **A–G** for each of the molecules or polyatomic ions listed on the report form. Refer back to the previous discussion, organized into corresponding sections A–G, for help with each step if necessary.

A. Number of Valence Electrons in a Molecule or Polyatomic Ion

Use a periodic table to determine the number of valence electrons for each group of atoms in the first column of the report form.

example: SiF_4 Si is in Group 4A, it has 4 valence electrons
F is in Group 7A, it has 7 valence electrons

Total valence electrons is $4 + 4(7) = 32$ electrons

If the group is a polyatomic ion, total the electrons as above, then add one electron for each negative charge or subtract one electron for each positive charge.

example: CO_3^{2-} C is in group 4A, it has 4 valence electrons
O is in Group 6A, it has 6 valence electrons
Ion has a −2 charge, add 2 electrons

Total valence electrons is $4 + 3(6) + 2 = 24$ electrons

B. Lewis Structures for Molecules and Polyatomic Ions

Use the following rules to show the two-dimensional Lewis structure for each molecule or polyatomic ion. Put your structure in the space provided. Use a ***sharp*** pencil and be as neat as possible.

1. Write down the skeletal arrangement of the atoms and connect them with a single covalent bond (a short line). We want to keep the rules at a minimum for this step, but we also want to avoid arrangements which will later prove incorrect. Useful guidelines are

    a. carbon is usually a central atom or forms bonds with itself; if carbon is absent, the central atom is usually the least electronegative atom in the group;

    b. hydrogen, which has only one valence electron, can form only one covalent bond and is never a central atom;

    c. oxygen atoms are not normally bonded to each other except in peroxides, and oxygen atoms normally have a maximum of two covalent bonds (two single bonds or one double bond).

Using these guidelines, skeletal arrangements for SiF_4 and CO_3^{2-} are

```
    F
    |
  F–Si–F          O–C–O
    |               |
    F               O
```

2. Subtract two electrons from the total valence electrons for each single bond used in the skeletal arrangement. This calculation gives the net number of electrons available for completing the electron structure. In the examples above, there are 4 and 3 single bonds, respectively. With 2 e^- per bond the calculation is

SiF_4: $32\,e^- - 4(2\,e^-) = 24\,e^-$ left to be assigned to the molecule

CO_3^{2-}: $24\,e^- - 3(2\,e^-) = 18\,e^-$ left to be assigned to the polyatomic ion

3. Distribute these remaining electrons as pairs of dots around each atom (except hydrogen) to give each atom a total of eight electrons around it. If there are not enough electrons available, move on to step 4.

```
       ..
      :F:
   ..  |  ..                ..       ..
  :F–Si–F:                 :O–C–O:
   ..  |  ..                ..  |  ..
      :F:                      :O:
       ..                       ..
```

all atoms have 8 electrons so the Lewis structure is complete

C does not have an octet of electrons so it is necessary to continue on with step 4

4. Check each Lewis structure to determine if every atom except hydrogen has an octet of electrons. If there are not enough electrons to give each of these atoms eight electrons, change single bonds between atoms to double or triple bonds by shifting unshared pairs of electrons as needed. A double bond counts as $4\,e^-$ for each atom to which it is bonded.

For CO_3^{2-}, shift 2 e^- from one of the O atoms and place it between C and that O.

Now, all the atoms have $8e^-$ around them. (Don't forget the $^{2-}$)

C. Model Building

1. Use the balls and sticks from the kit provided to build a 3-dimensional model of the molecule or polyatomic ion for each Lewis structure in the report form.

   a. Use a ball with 4 holes for the central atom.

   b. Use inflexible sticks for single bonds.

   c. Use flexible connectors for double or triple bonds.

   d. Use inflexible sticks for lone pairs around the central atom only.

2. **Leave the model together until it is checked by the instructor.** If you have to wait for someone to check your model, start building the next model on the list. If you complete each structure so fast that you run out of components before someone checks your models, work on other parts of the experiment.

D. Molecular Geometry

Look at your model from all angles and compare its structure to the description in the discussion (Section D). Then identify its molecular geometry from the following list and write the name of the geometric pattern on the report form in column D.

1. tetrahedral
2. trigonal pyramidal
3. trigonal planar
4. bent
5. linear

E. Central Bond Angles

Fill in column E with the bond angles between the central atom and all atoms attached to it. Review the discussion (Section E) to find the value of the angles associated with each geometric form. For molecules with more than one central atom, give bond angles for each. For molecules without a central atom and hence no bond angle, write *no central atom.*

F. Bond Polarity

Bond polarity can be determined by looking up the electronegativity values for both atoms in the Electronegativity table. In the F column of the report form, draw the symbols for both atoms involved in a bond and connect them with an arrow pointing toward the more electronegative atom. If there are several identical bonds it is only necessary to draw one. Use the following as examples.

N←+H   S+→O

G. Molecular Dipoles

Look at the model and evaluate its symmetry. Decide if the polar bonds within it cancel each other around the central atom resulting in a nonpolar molecule or if they do not cancel one another and result in a dipole. Some examples:

```
       :C̈l:                      H
        ↑                         ↓
:C̈l←+Si+→C̈l:              H+→C+→C̈l:
        ↓                         ↑
       :C̈l:                      H
```

symmetrical
nonpolar

asymmetrical
a dipole

Remember, it is also possible for all the polar bonds within a polyatomic ion to cancel each other so the resultant effect is nonpolar even though the group as a whole has a net charge.

```
⎡O↖          ⎤2–
⎢    C +⇒ O  ⎥
⎣O↙          ⎦
```

symmetrical
not a dipole

NAME Ellen Jaggers

SECTION ________ DATE 10-14-14

INSTRUCTOR Prof. Jaggi

REPORT FOR EXPERIMENT 17

Lewis Structures and Molecular Models

For each of the following molecules or polyatomic ions, fill out columns A through G using the instructions provided in the procedure section. These instructions are summarized briefly below.

A. Calculate the total number of valence electrons in each formula.
B. Draw a Lewis structure for the molecule or ion which satisfies the rules provided in the procedure.
C. Build a model of the molecule and have it checked by the instructor.
D. Use your model to determine the molecular geometry for this molecule (don't try to guess the geometry without the model): tetrahedral, trigonal pyramidal, trigonal planar, bent, linear
E. Determine the bond angle between the central atom and the atoms bonded to it. If there are only two atoms in the structure write "no central atom" in the space provided.
F. Use the electronegativity table to determine the electronegativity of the bonded atoms.
If the bonds are polar, indicate this with a modified arrow (⟷) pointing to the more electronegative element.
If the bonds are nonpolar, indicate this with a short line (—).
If there are two or more different atoms bonded to the central atom, include each bond.
G. Use your model and your knowledge of the bond polarity to determine if the molecule as a whole is nonpolar or a dipole.
If it is polar, write *dipole* in G. If it is not, write *nonpolar*.

	A	B	C	D	E	F	G
Molecule or Polyatomic Ion	**No. of Valence Electrons**	**Lewis Structure**		**Molecular Geometry**	**Bond Angles**	**Bond Polarity**	**Molecular Dipole or Nonpolar**
CH_4	8	H–C(–H)(–H)–H		tetrahedral	109.5°	H→C	nonpolar
CS_2	16	:S=C=S:		linear	180°	S←C→S	nonpolar

	A	B	C	D	E	F	G
Molecule or Polyatomic Ion	No. of Valence Electrons	Lewis Structure		Molecular Geometry	Bond Angles	Bond Polarity	Molecular Dipole or Nonpolar
H_2S	8	H–S–H (two lone pairs on S)		bent	104.5°	H→S	nonpolar
N_2	10	:N≡N:		linear	~~180°~~ none	N—N	nonpolar
SO_4^{2-}	14	[:S–O:]$^{-2}$		linear	none	S→O	~~non~~ polar (dipole)
H_3O^+	8	[H–O(–H)–H]$^+$		trigonal pyramidal	107°	H→O	nonpolar
CH_3Cl	14	H–C(H)(H)–Cl:		tetrahedral	109.5°	H→C C→Cl	polar (dipole)
C_2H_6	14	H–C(H)(H)–C(H)(H)–H		tetrahedral	109.5°	H→C	nonpolar
C_2H_4	12	H–C(H)=C(H)–H		trigonal planar	120°	H→C	non polar

	A	B	C	D	E	F	G
No. of Molecule or Polyatomic Ion	Valence Electrons	Lewis Structure		Molecular Geometry	Bond Angles	Bond Polarity	Molecular Dipole or Nonpolar
$C_2H_2Cl_2$	22	* Cl=C–C=Cl (H, H)		trigonal planar	120°	C←H C→Cl	dipole
SO_3^{2-}	~~24~~ 26	[O–S–O, O]$^{2-}$		trigonal pyramidal	107°	S→O	nonpolar
CH_2O	12	H–C(H)=O		trigonal planar	120°	C→O C←H	polar (dipole)
OF_2	20	F–O–F		Bent	104.5°	O→F	dipole
NO_2^-	18	[O–N=O]$^-$		Bent	104.5°	N→O	dipole
O_2	12	O=O		linear	none	O–O	nonpolar
NO_3^-	~~24~~ 24	** [O–N=O, O]$^-$		trigonal planar	120°	N→O	nonpolar

*More than one possible Lewis structure can be drawn. See questions 1, 2.
**More than one possible Lewis structure can be drawn. See question 3.

 NAME Ellen Jaggers

QUESTIONS

1. There are three acceptable Lewis structures for $C_2H_2Cl_2$ (*) and you have drawn one of them on the report form. Draw the other two structures and indicate whether each one is nonpolar or a dipole.

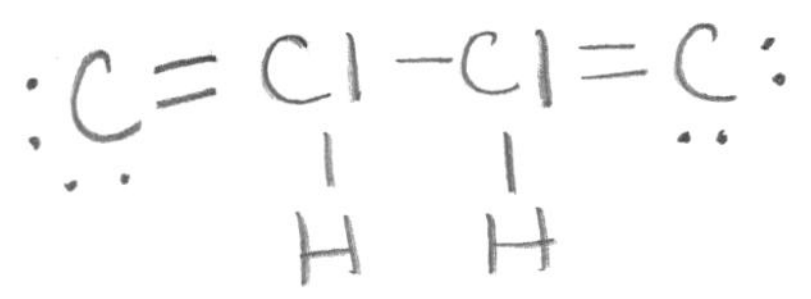

:C=Cl–Cl=C: (H on each Cl)

dipole

H–Cl–C≡C–Cl–H

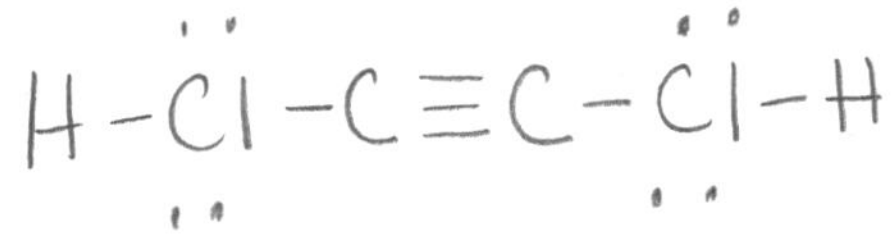

non polar

2. Explain why one of the three structures for $C_2H_2Cl_2$ is nonpolar and the other two are molecular dipoles.

The nonpolar structure has carbon tripled bonded to Carbon and is pulled equally on both sides by a Cl bonded to a H. It is linear. The dipoles are trigonal planar, with a more electronegative element on one side and H on the other, making the structure unbalanced, and therefore polar.

3. There are three Lewis structures for $[NO_3]^-$ (**). Draw the two structures which are not on the report form. Compare the molecular polarity of the three structures.

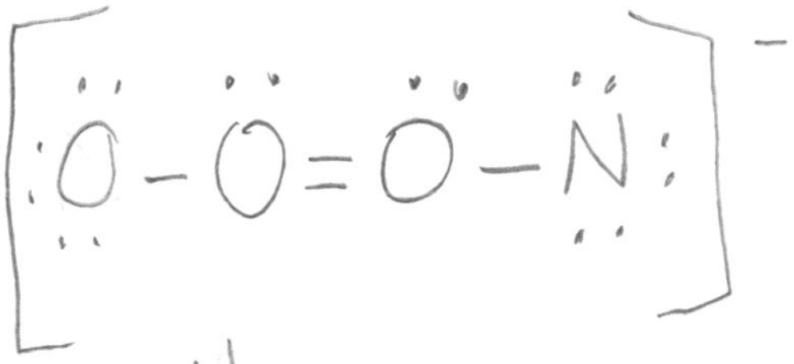

$[:\ddot{O}–\ddot{O}=\ddot{O}–\ddot{N}:]^-$

polar

$[:\ddot{O}–N=O–\ddot{O}:]^-$

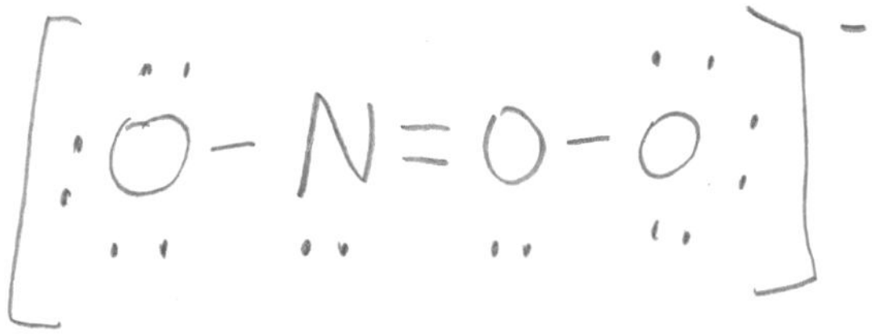

polar

The two above structures are polar, because there is greater electronegativity near the oxygens, making them unbalanced. The trigonal planar structure in the report is non polar, because the electronegativity is distributed around the nitrogen.

EXPERIMENT 22

Neutralization–Titration I

MATERIALS AND EQUIPMENT

Solid: potassium hydrogen phthalate, abbreviated KHP ($KHC_8H_4O_4$). **Liquids:** phenolphthalein indicator, unknown base solution (NaOH). One buret (25 mL or 50 mL) and buret clamp, buret brush. Wash bottle for distilled water.

DISCUSSION

The reaction of an acid and a base to form a salt and water is known as **neutralization.** In this experiment potassium hydrogen phthalate (abbreviated KHP) is used as the acid. Potassium hydrogen phthalate is an organic substance having the formula $HKC_8H_4O_4$, and like HCl, has only one acid hydrogen atom per molecule. Because of its complex formula, potassium hydrogen phthalate is commonly called KHP Despite its complex formula we see that the reaction of KHP with sodium hydroxide is similar to that of HCl. One mole of KHP reacts with one mole of NaOH.

$$HKC_8H_4O_4 + NaOH \longrightarrow NaKC_8H_4O_4 + H_2O$$

$$HCl + NaOH \longrightarrow NaCl + H_2O$$

Titration is the process of measuring the volume of one reagent required to react with a measured volume or mass of another reagent. In this experiment we will determine the molarity of a base (NaOH) solution from data obtained by titrating KHP with the base solution. The base solution is added from a buret to a flask containing a weighed sample of KHP dissolved in water. From the mass of KHP used we calculate the moles of KHP. Exactly the same number of moles of base is needed to neutralize this number of moles of KHP since one mole of NaOH reacts with one mole of KHP. We then calculate the molarity of the base solution from the titration volume and the number of moles of NaOH in that volume.

In the titration, the point of neutralization, called the **end-point,** is observed when an indicator, placed in the solution being titrated, changes color. The indicator selected is one that changes color when the stoichiometric quantity of base (according to the chemical equation) has been added to the acid. A solution of phenolphthalein, an organic acid, is used as the indicator in this experiment. Phenolphthalein is colorless in acid solution but changes to pink when the solution becomes slightly alkaline. When the number of moles of sodium hydroxide added is equal to the number of moles of KHP originally present, the reaction is complete. The next drop of sodium hydroxide added changes the indicator from colorless to pink.

Use the following relationships in your calculations:

1. According to the equation for the reaction,

   Moles of KHP reacted = Moles of NaOH reacted

2. $$\text{Moles} = \frac{\text{g of solute}}{\text{molar mass of solute}}$$

3. Molarity is an expression of concentration, the units of which are moles of solute per liter of solution:

$$\text{Molarity} = \frac{\text{moles}}{\text{liter}}$$

Thus, a 1.00 molar (1.00 M) solution contains 1.00 mole of solute in 1 liter of solution. A 0.100 M solution, then, contains 0.100 mole of solute in 1 liter of solution.

4. The number of moles of solute present in a known volume of solution of known concentration can be calculated by multiplying the volume of the solution (in liters) by the molarity of the solution:

$$\text{Moles} = (\text{liters})(\text{molarity}) = (\text{liters})\left(\frac{\text{moles}}{\text{liter}}\right)$$

PROCEDURE

Wear protective glasses.

Dispose of all solutions in the sink.

Make all weighings to the highest precision of the balance.

Obtain some solid KHP in a test tube or vial. Weigh two samples of KHP into 125 mL Erlenmeyer flasks, numbered for identification. (The flasks should be rinsed with distilled water, but need not be dry on the inside.) First weigh the flask, then add KHP to the flask by tapping the test tube or vial until 1.000 to 1.200 g has been added (see Figure 22.1). Determine the mass of the flask and the KHP. In a similar manner weigh another sample of KHP into the second flask. To each flask add approximately 30 mL of distilled water. If some KHP is sticking to the walls of the flask, rinse it down with water from a wash bottle. Warm the flasks slightly and swirl them until all the KHP is dissolved.

Figure 22.1 Method of adding KHP from a vial to a weighed Erlenmeyer flask

Obtain one buret and clean it. See "Use of the Buret," on the following page for instructions on cleaning and using the buret. Read and record all buret volumes to the nearest 0.01 mL.

Obtain about 250 mL of a base (NaOH) of unknown molarity in a clean, **dry** 250 mL Erlenmeyer flask as directed by your instructor. Record the number of this unknown.

1. Keep your base solution stoppered when not in use.
2. The 250 mL sample of base is intended to be used in both this experiment and Experiment 23. Be sure to label and save it.

Rinse the buret with two 5 to 10 mL portions of the base, running the second rinsing through the buret tip. Discard the rinsings in the sink. Fill the buret with the base, making sure that the tip is completely filled and contains no air bubbles. Adjust the level of the liquid in the buret so that the bottom of the meniscus is at exactly 0.00 mL. Record the initial buret reading (0.00 mL) in the space provided on the report form.

Add 3 drops of phenolphthalein solution to each 125 mL flask containing KHP and water. Place the first (Sample 1) on a piece of white paper under the buret extending the tip of the buret into the flask (see Figure 22.2).

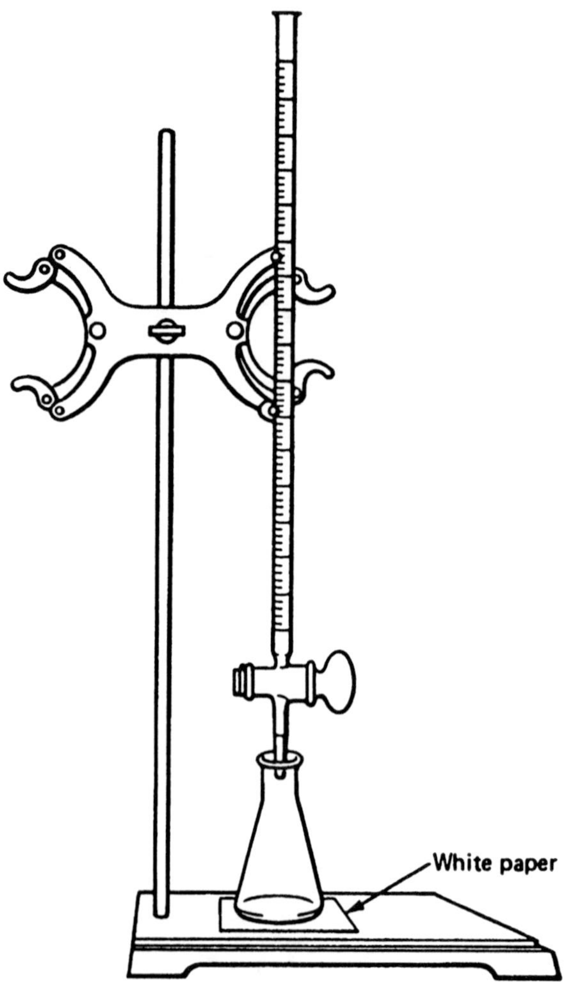

Figure 22.2 Setup with stopcock buret

Titrate the KHP by adding base until the end-point is reached. The titration is conducted by swirling the solution in the flask with the right hand (if you are right handed) while manipulating the stopcock with the left (Figure 22.3). As base is added you will observe a pink color caused by localized high base concentration. Toward the end-point the color flashes throughout the solution, remaining for a longer time. When this occurs, add the base drop by drop until the end-point is reached, as indicated by the first drop of base which causes a faint pink color to remain in the entire solution for at least 30 seconds. Read and record the final buret reading (see Figure 22.5). Refill the buret to the zero mark and repeat the titration with Sample 2. Then, calculate the molarity of the base in each sample. If these molarities differ by more than 0.004, titrate a third sample.

When you are finished with the titrations, empty and rinse the buret at least twice (including the tip) with tap water and once with distilled water. Return the vial with the unused KHP.

Use of the Buret

A buret is a volumetric instrument that is calibrated to deliver a measured volume of solution. The 50 mL buret is calibrated from 0 to 50 mL in 0.1 mL increments and is read to the nearest 0.01 mL. All volumes delivered from the buret should be between the calibration marks. (Do not estimate above the 0 mL mark or below the 50 mL mark.)

1. **Cleaning the Buret.** The buret must be clean in order to deliver the calibrated volume. Drops of liquid clinging to the sides as the buret is drained are evidence of a dirty buret.

To clean the buret, first rinse it a couple of times with tap water, pouring the water from a beaker. Then scrub it with a detergent solution, using a long-handled buret brush. Rinse the buret several times with tap water and finally with distilled water. Check for cleanliness by draining the distilled water through the tip and observe whether droplets of water remain on the inner walls of the buret.

2. **Using the Buret.** After draining the distilled water, rinse the buret with two 5 to 10 mL portions of the titrating solution to be used in it. This rinsing is done by holding the buret in a horizontal position and rolling the solution around to wet the entire inner surface. Allow the final rinsing to drain through the tip.

Fill the buret with the solution to slightly above the 0 mL mark and adjust it to 0.00 mL, or some other volume below this mark, by draining the solution through the tip. The buret tip must be completely filled to deliver the volume measured.

To deliver the solution from the buret, turn the stopcock with the forefinger and the thumb of your left hand (if you are right handed) to allow the solution to enter the flask. (See Figure 22.3). This procedure leaves your right hand free to swirl the solution in the flask during the titration. With a little practice you can control the flow so that increments as small as 1 drop of solution can be delivered.

3. **Reading the Buret.** The smallest calibration mark of a 50 mL buret is 0.1 mL. However, the buret is read to the nearest 0.01 mL by estimating between the calibration marks. When reading the buret be sure your line of sight is level with the bottom of the meniscus in order to avoid parallax errors (see Figure 22.4). The exact bottom of the meniscus may be made more prominent and easier to read by allowing the meniscus to pick up the reflection from a heavy dark line on a piece of paper (see Figure 22.5).

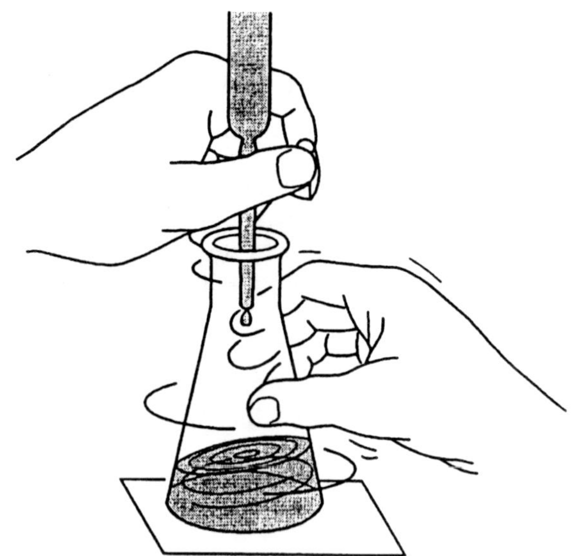

Figure 22.3 Titration technique

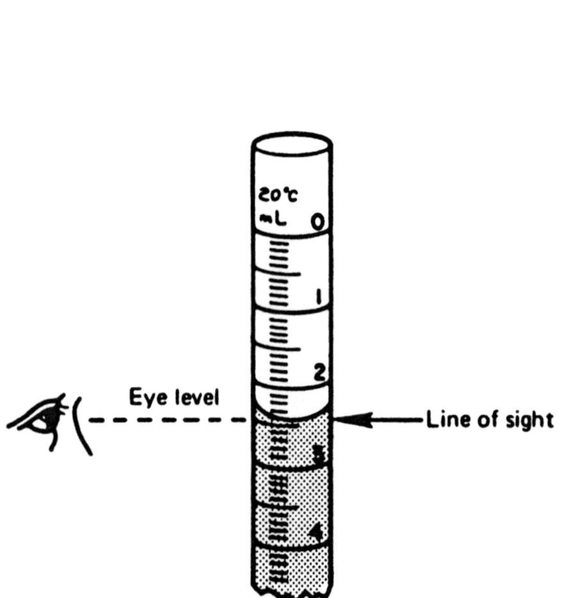

Figure 22.4 Reading the buret. The line of sight must be level with the bottom of the meniscus to avoid parallax.

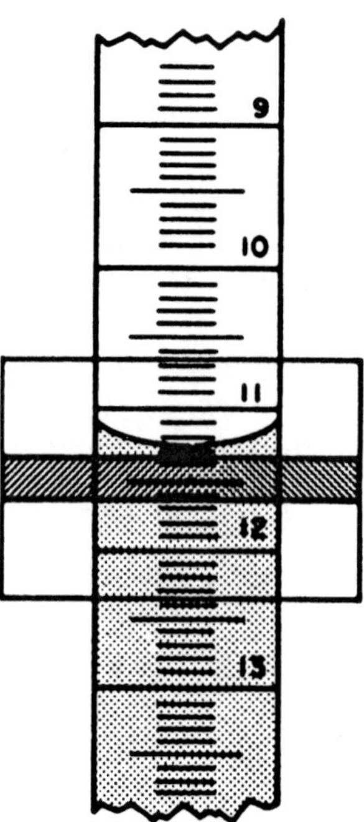

Figure 22.5 Reading the meniscus. A heavy dark line brought to within one division of the meniscus will make the meniscus more prominent and easier to read. The volume reading is 11.28 mL.

NAME Ellen Jaggers

SECTION ______ DATE ______

REPORT FOR EXPERIMENT 22

INSTRUCTOR Jaggi

Neutralization – Titration I

Data Table

	Sample 1	Sample 2	Sample 3 (if needed)
Mass of flask and KHP	69.29g	68.80g	
Mass of empty flask	68.26g	67.79g	
Mass of KHP	1.03g	1.01g	
Final buret reading	16.77 mL	32.18 mL	
Initial buret reading	0 mL	16.77 mL	
Volume of base used	16.77 mL	15.41 mL	

CALCULATIONS: In the spaces below show calculation setups **for Sample 1 only.** Show answers for both samples in the boxes. Remember to use the proper number of significant figures in all calculations. (The number 0.005 has only one significant figure.)

	Sample 1	Sample 2	Sample 3 (if needed)
1. Moles of acid (KHP, Molar mass = 204.2)	5.0×10^{-3} moles	4.9×10^{-3} moles	
2. Moles of base used to neutralize (react with) the above number of moles of acid	5.0×10^{-3} moles	4.9×10^{-3} moles	
3. Molarity of base (NaOH)	M=.298	M=.321	

molar mass = $\frac{g}{moles}$ 1.03/204.2 0.005 moles

$M = \frac{moles}{L}$ $\frac{5.0 \times 10^{-3}}{.01677}$

4. Average molarity of base .3095

5. Unknown base number sample# 4

QUESTIONS AND PROBLEMS

1. If you had added 50 mL of water to a sample of KHP instead of 30 mL, would the titration of that sample then have required more, less, or the same amount of base? Explain.

Adding 50 mL H_2O would lower the molarity of the acid, which would in turn lower the molarity of the base required to neutralize it.

2. A student weighed out 1.106 g of KHP How many moles was that?

$1.106\,g\,KHP \times \frac{1\,mol}{204.2\,g\,KHP}$

.0054 mol

3. A titration required 18.38 mL of 0.1574 M NaOH solution. How many moles of NaOH were in this volume?

moles = L(M)

.0029 mol

4. A student weighed a sample of KHP and found it weighed 1.276 g. Titration of this KHP required 19.84 mL of base (NaOH). Calculate the molarity of the base.

$M = \frac{moles}{L}$ $\frac{1.276\,g\,KHP}{204.2\,g/mol\,KHP} = .00625$ moles KHP

$M = \frac{.00625}{.01984\,L}$

0.315 M

5. Forgetful Freddy weighed his KHP sample, but forgot to bring his report sheet along, so he recorded the mass of KHP on a paper towel. During his titration, which required 18.46 mL of base, he spilled some base on his hands. He remembered to wash his hands, but forgot about the data on the towel, and used it to dry his hands. When he went to calculate the molarity of his base, Freddy discovered that he didn't have the mass of his KHP. His kindhearted instructor told Freddy that his base was 0.2987 M. Calculate the mass of Freddy's KHP sample.

M(L) = moles

$.01846(.2987) = .0055$ moles $\times \frac{204.2\,g}{1\,mol}$

1.126 g

6. What mass of solid NaOH would be needed to make 645 mL of Freddy's NaOH solution?

M(L) = moles

39.34 g

$.2987(.645) = .19266$ moles

$.19266$ moles $\times \frac{204.2\,g}{1\,mol}$

EXPERIMENT 23

Neutralization – Titration II

MATERIALS AND EQUIPMENT

Solutions: Acid of unknown molarity, standard base solution (NaOH), vinegar, phenolphthalein indicator. Suction bulb, buret, buret brush, buret clamp, 10 mL volumetric pipet. Wash bottle for distilled water.

DISCUSSION

This experiment may follow Experiment 22 or it may be completed independently of Experiment 22. In either case the discussion section of Experiment 22 supplements the following discussion.

The reaction of an acid and a base to form water and a salt is known as **neutralization.** Hydrochloric acid and sodium hydroxide, for example, react to form sodium chloride and water.

$$HCl(aq) + NaOH(aq) \longrightarrow H_2O(l) + NaCl(aq)$$

The ionic reaction in neutralizations of this type is that of hydrogen (or hydronium) ion reacting with hydroxide ion to form water.

$$H^+(aq) + OH^-(aq) \longrightarrow H_2O(l) \quad \text{or} \quad H_3O^+(aq) + OH^-(aq) \longrightarrow 2\,H_2O(l)$$

A monoprotic acid—i.e., an acid having one ionizable hydrogen atom per molecule—reacts with sodium hydroxide (or any other monohydroxy base) on a 1:1 mole basis. This fact is often utilized in determining the concentrations of solutions of acids by titration.

Titration is the process of measuring the volume of one reagent to react with a measured volume or mass of another reagent. In this experiment an acid solution of unknown concentration is titrated with a base solution of known concentration, Phenolphthalein is used as an indicator. This substance is colorless in acid solution, but changes to pink when the solution becomes slightly basic or alkaline. The change of color, caused by a single drop of the base solution in excess over that required to neutralize the acid, marks the **end-point** of the titration.

Molarity (M) is the concentration of a solution expressed in terms of moles of solute per liter of solution.

$$\text{Molarity} = \frac{\text{moles}}{\text{liter}}$$

Thus a solution containing 1.00 mole of solute in 1.00 liter of solution is 1.00 molar (1.00 M). If only 0.155 mole is present in 1.00 liter of solution, it is 0.155 M, etc. To determine the molarity of any quantity it is only necessary to divide the total number of moles of solute present in the solution by the volume (in liters).

To determine the number of moles of solute present in a known volume of solution, multiply the volume in liters by the molarity.

$$\text{Moles} = (\text{liters})(\text{molarity}) = (\text{liters})\left(\frac{\text{moles}}{\text{liter}}\right)$$

For titrations involving monoprotic acids and monohydroxy bases (one hydroxide ion per formula unit), the number of moles of acid is identical to the number of moles of base required to neutralize the acid. In this experiment we measure the volume of base of known molarity required to neutralize a measured volume of acid of unknown molarity. The molarity of the acid can then be calculated.

$$\text{Moles base} = (\text{liters})(\text{molarity}) = (\text{liters base})\left(\frac{\text{moles base}}{\text{liters}}\right)$$

$$\text{Moles acid} = (\text{moles base})\left(\frac{1\ \text{mole acid}}{1\ \text{mole base}}\right)$$

$$\text{Molarity of acid} = \frac{\text{moles acid}}{\text{liters acid}}$$

In order to determine the molarity of an acid solution, it is not actually necessary to know what the acid is—only whether it is monoprotic, diprotic, or triprotic. The calculations in this experiment are based on the assumption that the acid in the unknown is monoprotic.

If the molarity and the formula of the solute are known, the concentration in grams of solute per liter of the solution may be calculated by multiplying by the molar mass.

$$(\text{Molarity})(\text{molar mass}) = \left(\frac{\text{moles}}{\text{liter}}\right)\left(\frac{\text{grams}}{\text{mole}}\right) = \frac{\text{grams}}{\text{liter}}$$

In determining the acid content of commercial vinegar, it is customary to treat the vinegar as a dilute solution of acetic acid, $HC_2H_3O_2$. The acetic acid concentration of the vinegar may be calculated as grams of acetic acid per liter or as percent acid by mass. If the acetic acid content is to be expressed on a mass percent basis, the density of the vinegar must also be known.

PROCEDURE

Wear protective glasses.

Do not pipet by mouth.

Dispose of all solutions in the sink. Flush with water.

A. Molarity of an Unknown Acid

Obtain a sample of acid of unknown molarity in a clean, dry 125 mL Erlenmeyer flask as directed by your instructor.

With a volumetric pipet, transfer a 10.00 mL sample of the acid to a clean, but not necessarily dry, Erlenmeyer flask. See "Use of the Pipet," on the following page, for instructions on cleaning and using the pipet. Pipet a duplicate 10.00 mL sample into a second flask. (If pipets are not available, a buret which has been carefully cleaned and rinsed may be used to measure the acid samples.)

You will need about 150 mL of base of known molarity (standard solution). Your instructor will give you the exact molarity of the base solution that you used in Experiment 22 or you may be given another sodium hydroxide solution of known molarity. Record the exact molarity of this solution. Keep the flask containing the base stoppered when not in use.

Clean and set up a buret. See "Use of the Buret," in Experiment 22, for instructions on cleaning and using the buret.

Rinse the buret with two 5 to 10 mL portions of the base, running the second rinsing through the buret tip. Discard the rinsings in the sink. Fill the buret with the base, making sure that the tip is completely filled and contains no air bubbles. Adjust the level of the liquid in the buret so that the bottom of the meniscus is near or exactly at 0.00 mL. Record the initial buret reading in the space provided on the report form.

Add three drops of phenolphthalein solution and about 25 mL of distilled water to the flask containing the 10.00 mL of acid. Place this flask on a piece of white paper under the buret and lower the buret tip into the flask (see Figure 22.2).

Titrate the acid by adding base until the end-point is reached. During the titration swirl the solution in the flask with the right hand (if you are right handed) while manipulating the stopcock with the left. As the base is added you will observe a pink color caused by localized high base concentration. Near the end-point this color flashes throughout the solution, remaining for increasingly longer periods of time. When this occurs, add the base drop by drop until the end-point is reached, as indicated by the first drop of base which causes the entire solution to retain a faint pink color for at least 30 seconds. Record the final buret reading.

Refill the buret with base and adjust the volume to near the zero mark. Titrate the duplicate sample of acid. If the volumes of base used differ by more than 0.20 mL, titrate a third sample. In the calculations, assume that the unknown acid reacts like KHP or HCl (one mole of acid reacts with one mole of base).

B. Acetic Acid Content of Vinegar

Obtain about 40 mL of vinegar in a clean, dry 50 mL beaker. Record the sample number, if any, of this vinegar.

Titrate duplicate 10.00 mL samples of vinegar using exactly the same procedure outlined in Part A. Remember to rinse the pipet with vinegar before pipeting the vinegar samples.

When you are finished with the titrations, empty the buret and rinse it and the pipet at least twice with tap water and once with distilled water.

Use of the Pipet

A volumetric (transfer) pipet (Figure 23.1) is calibrated to deliver a specified volume of liquid to a precision of about ±0.02 mL in a 10 mL pipet. To achieve this precision, the pipet must be clean and used in a specified manner.

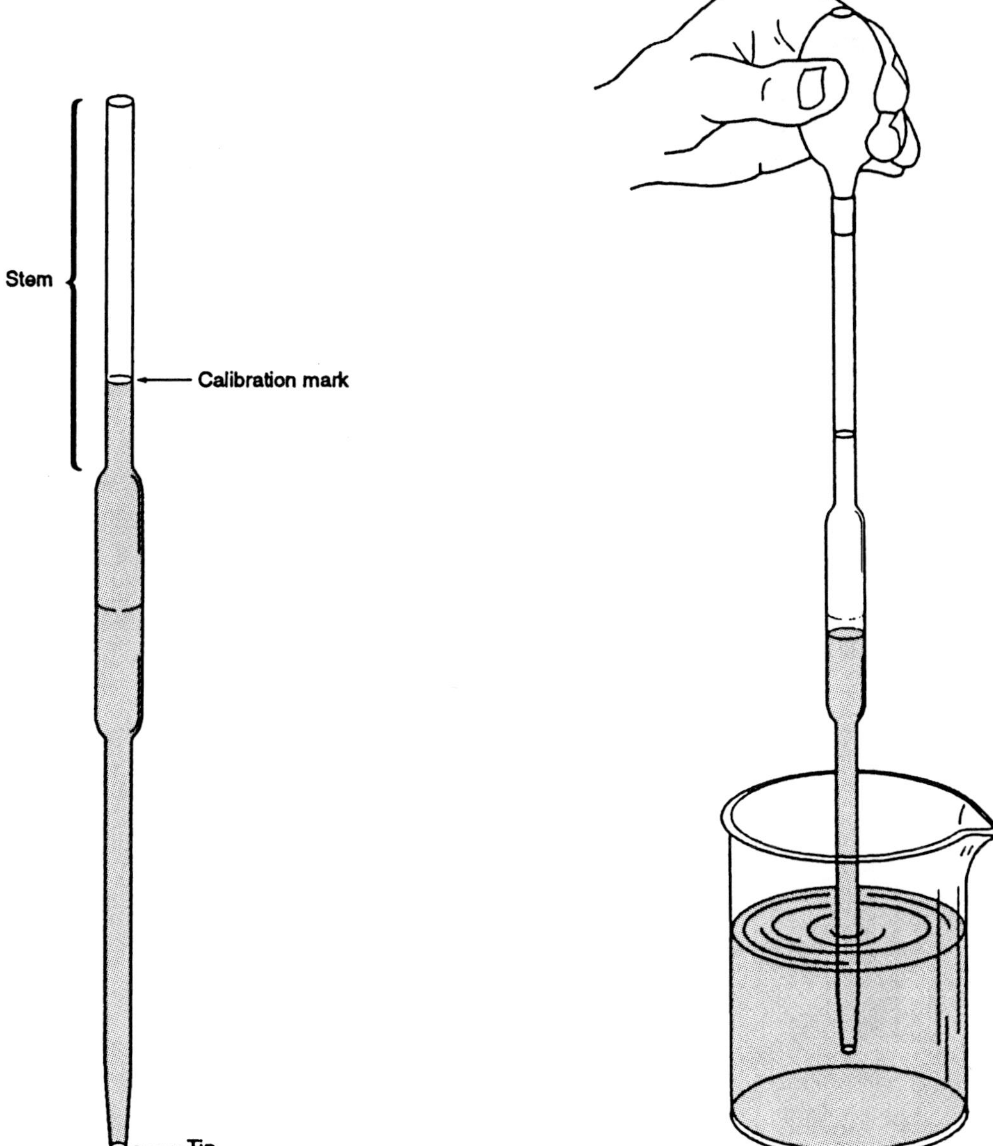

Figure 23.1 A volumetric (transfer) pipet

Figure 23.2 Liquid is drawn into the pipet with a rubber suction bulb. Keep the tip of the pipet below the liquid level during suction.

Liquids are drawn into a pipet by means of a rubber suction bulb (Figure 23.2) or by a rubber tube connected to a water aspirator pump. Suction by mouth has also been used to draw liquids into a pipet, but this is a dangerous practice and is not recommended.

1. **Cleaning the Pipet.** Use a rubber suction bulb to draw up enough detergent solution to fill about two-thirds of the body or bulb of the pipet. Retain this solution in the pipet by pressing the forefinger tightly against the top of the pipet stem (Figure 23.3). turn the pipet to a nearly horizontal position and gently shake and rotate it until the entire inside surface is wetted. Allow the pipet to drain and rinse it at least three times with tap water and once with distilled water.

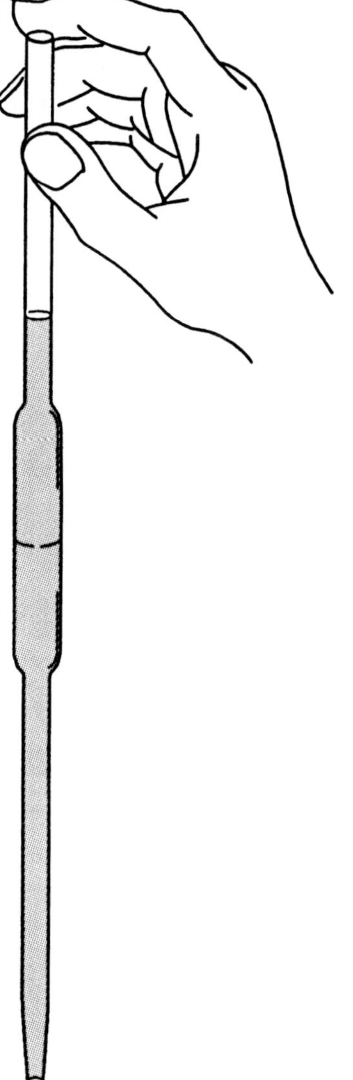

Figure 23.3 Liquid is retained in the pipet by applying pressure with the forefinger to the top of the stem.

Figure 23.4 The pipet is calibrated to deliver the specified volume, leaving a small amount of liquid in the tip.

2. **Using the Pipet.** Unless the pipet is known to be clean and absolutely dry on the inside, it must be rinsed twice with small portions of the liquid that is to be pipeted. This is done as in the washing procedure described above. These rinses are discarded in order to avoid contamination of the liquid being pipeted. A **pipet** does not need to be rinsed between successive pipettings of the same solution.

To transfer a measured volume of a liquid, collapse a suction bulb by squeezing and place it tightly against the top of a pipet. (Do not try to push the bulb on to the pipet.) Draw the liquid into the pipet until it is filled to about 5 cm above the calibration mark by allowing the bulb to slowly expand. Be careful—do not allow the liquid to get into the bulb. Remove the bulb and quickly place your forefinger over the top of the pipet stem. The liquid will be retained in the pipet if the finger is pressed tightly against the top of the stem. Keeping the pipet in a vertical position, decrease the finger pressure very slightly, and allow the liquid level to drop slowly toward the calibration mark. When the liquid level has almost reached the calibration mark, again increase the finger pressure and stop the liquid when the bottom of the meniscus is exactly on the calibration mark. Touch the tip to the wall of the flask to remove the adhering drop of liquid.

Move the pipet to the flask which is to receive the sample and allow the liquid to drain while holding the pipet in a vertical position. About 10 seconds after the liquid has stopped running from the pipet, touch the tip to the inner wall of the sample flask to remove the drop of liquid adhering to the tip. A small amount of liquid will remain in the tip (Figure 23.4). Do not blow or shake this liquid into the sample; the pipet is calibrated to deliver the volume specified without this small residual.

If you have never used a volumetric pipet, it is advisable to practice by pipetting some samples of distilled water until you have mastered the technique.

NAME ______________________

SECTION ______ DATE ______

REPORT FOR EXPERIMENT 23

INSTRUCTOR______________

Neutralization-Titration II

A. Molarity of an Unknown Acid

Data Table

	Sample 1		**Sample 2**		**Sample 3 (if needed)**	
	Acid*	Base	Acid*	Base	Acid*	Base
Final buret reading						
Initial buret reading						
Volume used						

*If a pipet is used to measure the volume of acid, record only in the space for volume used.

Molarity of base (NaOH) ________________

CALCULATIONS: In the spaces below, show calculation setups for Sample 1 only. Show answers for both samples in the boxes.

	Sample 1	**Sample 2**	**Sample 3 (if needed)**
1. Moles of base (NaOH) (if needed)			
2. Moles of acid used to neutralize (react with) the above number of moles of base			
3. Molarity of acid			

4. Average molarity of acid ______________________

5. Unknown acid number ______________________

B. Acetic Acid Content of Vinegar

Data Table

	Sample 1		**Sample 2**		**Sample 3 (if needed)**	
	Vinegar*	Base	Vinegar*	Base	Vinegar*	Base
Final buret reading						
Initial buret reading						
Volume used						

*If a pipet is used to measure the volume of vinegar, record only in the space for volume used.

Molarity of base (NaOH) ________________ Vinegar number ________________

CALCULATIONS: In the spaces below, show calculation setups for Sample 1 only. Show answers for both samples in the boxes.

	Sample 1	**Sample 2**	**Sample 3 (if needed)**
1. Moles of base (NaOH)			
2. Moles of acid ($HC_2H_3O_2$) used to neutralize (react with) the above number of moles of base			
3. Molarity of acetic acid in the vinegar			

4. Average molarity of acetic acid in the vinegar ________________

5. Grams of acetic acid per liter (from average molarity) ________________

6. Mass percent acetic acid in vinegar sample (density of vinegar = 1.005 g/mL) ________________

EXPERIMENT 24

Chemical Equilibrium – Reversible Reactions

MATERIALS AND EQUIPMENT

Solid: ammonium chloride (NH_4Cl). **Solutions:** saturated ammonium chloride, 0.1 M cobalt(II) chloride ($CoCl_2$), 0.1 M iron(III) chloride ($FeCl_3$), concentrated (12 M) hydrochloric acid (HCl), 0.1 M copper(II) sulfate ($CuSO_4$), 6 M ammonium hydroxide (NH_4OH), phenolphthalein, 0.1 M potassium thiocyanate (KSCN), 0.1 M silver nitrate ($AgNO_3$), saturated sodium chloride (NaCl), and dilute (3 M) sulfuric acid (H_2SO_4).

DISCUSSION

In many chemical reactions the reactants are not totally converted to the products because of a reverse reaction; that is, because the products react to form the original reactants. Such reactions are said to be reversible and are indicated by a double arrow ($\rightleftharpoons$) in the equation. The reaction proceeding to the right is called the **forward reaction;** that to the left, the **reverse reaction.** Both reactions occur simultaneously.

Every chemical reaction proceeds at a certain rate or speed. The rate of a reaction is variable and depends on the concentrations of the reactants and the conditions under which the reaction is conducted. When the rate of the forward reaction is equal to the rate of the reverse reaction, a condition of **chemical equilibrium** exists. At equilibrium the products react at the same rate as they are produced. Thus the concentrations of substances in equilibrium do not change, but both reactions, forward and reverse, are still occurring.

The principle of Le Chatelier relates to systems in equilibrium and states that when the conditions of a system in equilibrium are changed the system reacts to counteract the change and reestablish equilibrium. In this experiment we will observe the effect of changing the concentration of one or more substances in a chemical equilibrium. Consider the hypothetical equilibrium system

$$\mathbf{A + B \rightleftharpoons C + D}$$

When the concentration of any one of the species in this equilibrium is changed, the equilibrium is disturbed. Changes in the concentrations of all the other substances will occur to establish a new position of equilibrium. For example, when the concentration of B is increased, the rate of the forward reaction increases, the concentration of A decreases, and the concentrations of C and D increase. After a period of time the two rates will become equal and the system will again be in equilibrium. The following statements indicate how the equilibrium will shift when the concentrations of A, B, C, and D are changed.

An increase in the concentration of A or B causes the equilibrium to shift to the right.

An increase in the concentration of C or D causes the equilibrium to shift to the left.

A decrease in the concentration of A or B causes the equilibrium to shift to the left.

A decrease in the concentration of C or D causes the equilibrium to shift to the right.

Evidence of a shift in equilibrium by a change in concentration can easily be observed if one of the substances involved in the equilibrium is colored. The appearance of a precipitate or the change in color of an indicator can sometimes be used to detect a shift in equilibrium.

Net ionic equations for the equilibrium systems to be studied are given below. These equations will be useful for answering the questions in the report form.

A. Saturated Sodium Chloride Solution

$$NaCl(s) \overset{H_2O}{\rightleftharpoons} Na^+(aq) + Cl^-(aq)$$

$NaCl \xrightarrow{H_2O} Na^+(aq) + Cl^-(aq)$ inc

B. Saturated Ammonium Chloride Solution

$$NH_4Cl(s) \overset{H_2O}{\rightleftharpoons} NH_4^+(aq) + Cl^-(aq)$$

C. Iron(III) Chloride plus Potassium Thiocyanate

$$Fe^{3+}(aq) + KSCN^-(aq) \rightleftharpoons Fe(SCN)^{2+}(aq)$$

(+ $AgNO_3$)

$Fe^{3+}(aq)$	$SCN^-(aq)$	$Fe(SCN)^{2+}(aq)$
Pale yellow	Colorless	Red

D. Copper(II) Sulfate Solution with Ammonia

$$Cu(H_2O)_4^{2+}(aq) + 4\,NH_3(aq) \rightleftharpoons Cu(OH)_2(s) \rightleftharpoons [Cu(NH_3)_4]^{2+}(aq) + 4\,H_2O$$

$Cu(H_2O)_4^{2+}(aq)$	$Cu(OH)_2(s)$	$[Cu(NH_3)_4]^{2+}(aq)$
light blue clear	blue cloudy	deep blue/purple cloudy

E. Cobalt(II) Chloride Solution

The equilibrium involves the following ions in solutions:

$$Co(H_2O)_6^{2+}(aq) + 4\,Cl^-(aq) \rightleftharpoons CoCl_4^{2-}(aq) + 6\,H_2O(l)$$

$Co(H_2O)_6^{2+}(aq)$	$CoCl_4^{2-}(aq)$
Pink	Blue

F. Ammonia Solution

$$NH_3(aq) + H_2O(l) \rightleftharpoons NH_4^+(aq) + OH^-(aq)$$

PROCEDURE

Wear protective glasses.

> **NOTE:** Record observed evidence of equilibrium shifts as each experiment is done.

A. Saturated Sodium Chloride Solution

Add a few drops of conc. hydrochloric acid to 2 to 3 mL of saturated sodium chloride solution in a test tube, and note the results.

B. Saturated Ammonium Chloride Solution

Repeat Part A, using saturated ammonium chloride solution instead of sodium chloride solution.

Dispose of the solutions in A and B in the sink and flush with water.

C. Iron(III) Chloride plus Potassium Thiocyanate

Prepare a stock solution to be tested by adding 2 mL each of 0.1 M iron(III) chloride and 0.1 M potassium thiocyanate solutions to 100 mL of distilled water and mix. Pour about 5 mL of this stock solution into each of four test tubes.

1. Use the first tube as a control for color comparison.

2. Add about 1 mL of 0.1 M iron(III) chloride solution to the second tube and observe the color change.

3. Add about 1 mL of 0.1 M potassium thiocyanate solution to the third tube and observe the color change.

4. Add 0.1 M silver nitrate solution dropwise (less than 1 mL) to the fourth tube until almost all the color is discharged. The white precipitate formed consists of both AgCl and AgSCN. Pour about half the contents (including the precipitate) into another tube. Add 0.1 M potassium thiocyanate solution dropwise (1 to 2 mL) to one tube and 0.1 M iron(III) chloride solution (1 to 2 mL) to the other. Observe the results.

Dispose of the contents in tubes C.1–3 and the unused stock solutions in the sink and flush with water. Dispose of the contents of both C.4 tubes in the "heavy metals" waste container.

D. Copper (II) Sulfate Solution with Ammonia

Pour 2 mL of 0.1 M copper (II) sulfate into each of two test tubes. Add 6 M $NH_3(aq)$ (NH_4OH) dropwise (shake well after each drop is added) to one of the copper(II) sulfate tubes. When there is a definite color or appearance change, note the change on the report form. Use the second test tube for comparison. Continue to add the $NH_3(aq)$ until there is another color or appearance change. Note the changes on the report form.

Now, add 3 M H_2SO_4 dropwise to the solution until the original color is restored. Again, use the second tube for comparison.

Dispose of the contents of both test tubes in the "heavy metals" waste container provided.

E. Cobalt(II) Chloride Solution

Place about 2 mL (no more) of 0.1 M cobalt(II) chloride solution into each of three test tubes.

1. To one tube add about 3 mL of conc. hydrochloric acid dropwise and note the result. Now add water dropwise to the solution until the original color (reverse reaction) is evident.

2. To the second tube add about 1.5 g of solid ammonium chloride and shake to make a saturated salt solution. Compare the color with the solution in the third tube (control). Place the second and third tubes (unstoppered) in a beaker of boiling water, shake occasionally, and note the results. Cool both tubes under tap water until the original color (reverse reaction) is evident.

Dispose of these solutions in the "heavy metals" waste container.

F. Ammonia Solution

Prepare an ammonia stock solution by adding 10 drops of 6 M ammonium hydroxide and 3 drops of phenolphthalein to 100 mL of tap water and mix. Pour about 5 mL of this stock solution into each of two test tubes.

1. Dissolve a very small amount of solid ammonium chloride in the stock solution in the first tube and observe the result.

2. Add a few drops of dil. (6 M) hydrochloric acid to the stock solution in the second tube. Mix and observe the result.

Dispose of these solutions and the rest of the ammonia stock solution in the sink and flush with water.

NAME Ellen Jaggers

SECTION ______ DATE __________

REPORT FOR EXPERIMENT 24

INSTRUCTOR Prof Jaggi

Chemical Equilibrium – Reversible Reactions

Refer to equilibrium equations in the discussion when answering these questions.

A. Saturated Sodium Chloride

1. What is the evidence for a shift in equilibrium?

   A white precipitate formed.

2. Which ion caused the equilibrium to shift? Cl^-

3. In which direction did the equilibrium shift? to the left

4. If solid sodium hydroxide were added to neutralize the hydrochloric acid, would this reverse the reaction and cause the precipitated sodium chloride to redissolve? Explain.

   It would cause an increase in the product Na^+. If it brought the Na^+ and Cl^- back into equilibrium, then yes, it would cause the precipitate to redissolve.

B. Saturated Ammonium Chloride

1. What is the evidence for a shift in equilibrium?

   White precipitate formed.

2. In which direction did the equilibrium shift? Cl^-

3. Which ion caused the equilibrium to shift? to the left

C. Iron(III) Chloride plus Potassium Thiocyanate

1. What is the evidence for a shift in equilibrium when iron(III) chloride is added to the stock solution?

   The color changed to a darker orange.

2. What is the evidence for a shift in equilibrium when potassium thiocyanate is added to the stock solution?

   The color turned red.

 NAME Ellen Jaggers

3. (a) What is the evidence for a shift in equilibrium when silver nitrate is added to the stock solution? (The formation of a precipitate is not the evidence since the precipitate is not one of the substances in the equilibrium.)

Color changed to clear.

(b) The change in concentration of which ion in the equilibrium caused this equilibrium shift?

NO_3^-

(c) Write a net ionic equation to illustrate how this concentration change occurred.

KCl (aq) + Fe(CSN3)(aq) + AgNO3 → AgCl(s) + AgSCN(s) + KNO3(aq) + FeNO3(aq)

(d) When the mixture in C.4 was divided and further tested, what evidence showed that the mixture still contained Fe^{3+} ions in solution?

The color changed to peach.

D. Copper(II) Sulfate Solution with Ammonia

1. What was the evidence for the first shift in equilibrium when the $NH_3(aq)$ was added dropwise to the Cu^{2+} solution?

Color change – from sea blue to cloudy blue.

2. (a) Explain how adding more $NH_3(aq)$ caused the equilibria to shift again.

This created more $[Cu(NH_3)_4]^{2+}$, which is a deep blue color.

(b) What did you observe in the Cu^{2+} system to indicate that the shift had occured?

The color changed from cloudy blue to cobalt blue.

3. (a) Explain how 3 M sulfuric acid caused the equilibria to shift back again?

The H+ and O-2 concentrations increased, driving the reaction to the left.

(b) What did you observe to indicate that the reaction shifted to the left?

The color went back to it's original color.

E. Cobalt(II) Chloride Solution

1. What was the evidence for a shift in equilibrium when conc. hydrochloric acid was added to the cobalt chloride solution?

   color change from light pink to royal blue.

2. (a) Write the equilibrium equation for this system.

   $Co(H_2O)_6^{2+}(aq) + 4Cl^-(aq) \rightleftharpoons CoCl_4^{2-}(aq) + 6H_2O(l)$

   (b) State whether the concentration of each of the following substances was increased, decreased, or unaffected when the conc. hydrochloric acid was added to cobalt chloride solution.

   $Co(H_2O)_6^{2+}$ unaffected, Cl^- increased, $CoCl_4^{2-}$ increased

3. (a) What did you observe when ammonium chloride was added to cobalt chloride solution?

   pink with white on the bottom

   (b) What did you observe when this mixture was heated?

   Formation of a precipitate

   (c) Explain why heating the mixture caused the equilibrium to shift.

   It evaporated the H_2O, causing a precipitate to form.

   (d) What did you observe when the mixture was cooled?

   thick solid with white on botton

   3rd, same color as before

   (e) Explain why cooling the mixture caused the equilibrium to shift.

   It caused the aqueous substances to become solid.

 NAME Ellen Jaggers

F. Ammonia Solution

1. What is the evidence for a shift in equilibrium when ammonium chloride was added to the stock solution?

   Color turned from pink to clear.

2. Explain, in terms of the equilibrium, the results observed when hydrochloric acid was added to the stock solution. Color turned from pink to clear, indicating that the reaction favored the left.

3. State whether the concentration of each of the following was increased, decreased, or was unaffected when dilute hydrochloric acid was added to the ammonia stock solution:

NH_3 decreased, NH_4^+ decreased, OH^- increased, Pink color decreased

4. (a) In which direction would the equilibrium shift if sodium hydroxide were added to the ammonia stock solution? it would shift left.

   (b) Would the sodium hydroxide tend to decrease the color intensity? Explain.

   NO, it is not an acid, therefore it would not neutralize the basic NH_4OH. NaOH is also a base.

5. Would boiling the ammonia solution have any effect on the equilibrium? Explain.

   Boiling would evaporate the H_2O, which would cause the equilibrium to shift less, since the concentration of H_2O would be decreased.

MOLECULAR MASS OF AN IDEAL GAS

The molar mass (*MM* sometimes called the gram molar mass, or the molecular weight) of a gaseous compound can be calculated if the following are known: the mass, volume, temperature, and pressure of the gas. The ideal gas law PV = nRT allows one to calculate the number of moles of a gas if the other variables in this relationship are known. Rearranging the equation to express for the number of moles gives:

$$n = \frac{PV}{RT}$$

where n is the number of moles, P is the pressure in mm/Hg, V is volume in liters, T is the temperature in degrees Kelvin and R is the universal gas constant (62.4 mmHg*L/mol*K). The number of moles of any substance is related to its molecular mass *MM* by the following relationship:

$$n = \frac{\text{grams}}{MM}$$

In this experiment the molar mass (*MM*) of an unknown gas will be determined by weighing out a quantity of the gas (in grams) and calculating the molar mass (*MM*) by combining the two previous relationships into one final equation:

$$MM = \frac{\text{grams X RT}}{PV}$$

To determine the weight in grams of the unknown gas, a small amount of volatile liquid will be placed into a test tube. The test tube is covered with aluminum foil containing a small hole. This allows for the heated liquid to evaporate as its vapor is trapped inside the test tube. Since the complete apparatus is exposed to the environment, the pressure inside the test tube will equal the current atmospheric pressure. The vapor temperature will be that of the boiling water in which it is immersed. The volume of the gas can be calculated by calculating the volume of the test tube. The weight of the gas will be measured after rapidly cooling the vapor and weighing the condensed liquid remaining. All the variables needed to calculate the molar mas (*MM*) are therefore available.

PROCEDUE

CAUTION: The liquid used in this experiment is volatile and flammable! There should be no open flame anywhere near this experiment. Dispose of any remaining liquid in an appropriate container.

1. Prepare a boiling water bath by filling a 250 mL beaker half way with water and heating the contents on a hot plate. At the same time prepare a small ice bath using a similarly sized beaker.

2. Prepare a piece of aluminum foil so that it covers the top of a medium sized test tube.

3. Use a needle to make a small hole in the aluminum foil.

4. Record the weight of the test tube and aluminum foil in your data section.

5. Measure ~~0.5~~ 1.0 mL of the unknown liquid in a small graduated cylinder and pour the liquid into the test tube.

6. Immerse the covered test tube in the boiling water bath using test tube tongs. Ensure that the test tube remains submerged in the boiling water by holding with the tongs. The test tube only need be submerged so that the liquid remains below the water level. Avoid any steam from the boiling water condensing back into the test tube.

7. Record today's barometric pressure in mm/Hg on your data sheet. As the liquid continues to vaporize, the excess vapor will escape out of the pin hole. The vapor will fill the test tube at the temperature of the boiling water and at a pressure equal to the surrounding room pressure. Carefully monitor the amount of liquid remaining. Once all the liquid has vaporized continue to hold the test tube submerged for two more minutes. Measure and record the boiling water temperature (in both Celsius and Kelvin) in your data sheet.

8. Quickly cool the test tube in the ice bath, dry off the test tube and record the mass of the test tube, foil and condensed liquid.

9. Clean the test tube and completely fill it with water, covering the test tube with the same piece of aluminum foil. Use the density of water to calculate and record the volume of the test tube.

10. Your final molar mass (MM) calculation should be entered in your data sheet. Pay close attention to the significant figures in these calculations.